Springer Theses

Recognizing Outstanding Ph.D. Research

Aims and Scope

The series "Springer Theses" brings together a selection of the very best Ph.D. theses from around the world and across the physical sciences. Nominated and endorsed by two recognized specialists, each published volume has been selected for its scientific excellence and the high impact of its contents for the pertinent field of research. For greater accessibility to non-specialists, the published versions include an extended introduction, as well as a foreword by the student's supervisor explaining the special relevance of the work for the field. As a whole, the series will provide a valuable resource both for newcomers to the research fields described, and for other scientists seeking detailed background information on special questions. Finally, it provides an accredited documentation of the valuable contributions made by today's younger generation of scientists.

Theses are accepted into the series by invited nomination only and must fulfill all of the following criteria

- They must be written in good English.
- The topic should fall within the confines of Chemistry, Physics, Earth Sciences, Engineering and related interdisciplinary fields such as Materials, Nanoscience, Chemical Engineering, Complex Systems and Biophysics.
- The work reported in the thesis must represent a significant scientific advance.
- If the thesis includes previously published material, permission to reproduce this must be gained from the respective copyright holder.
- They must have been examined and passed during the 12 months prior to nomination.
- Each thesis should include a foreword by the supervisor outlining the significance of its content.
- The theses should have a clearly defined structure including an introduction accessible to scientists not expert in that particular field.

More information about this series at http://www.springer.com/series/8790

Adam Nahum

Critical Phenomena in Loop Models

Doctoral Thesis accepted by the University of Oxford, UK

Author
Dr. Adam Nahum
Department of Physics
Massachusetts Institute of Technology
Cambridge, MA
USA

Supervisor
Prof. John Chalker
Theoretical Physics
University of Oxford
Oxford
UK

ISSN 2190-5053 ISSN 2190-5061 (electronic)
ISBN 978-3-319-06406-2 ISBN 978-3-319-06407-9 (eBook)
DOI 10.1007/978-3-319-06407-9

Library of Congress Control Number: 2014949370

Springer Cham Heidelberg New York Dordrecht London

Printed on acid-free paper

Springer is part of Springer Science+Business Media (www.springer.com)

Publications Related to this Work

- A. Nahum, J. T. Chalker, P. Serna, M. Ortuño and A. M. Somoza, *3D Loop Models and the* CP^{n-1} *Sigma Model*, Phys. Rev. Lett. **107**, 110601 (2011).
- A. Nahum and J. T. Chalker, *Universal Statistics of Vortex Lines*, Phys. Rev. E **85**, 031141 (2012).
- A. Nahum, P. Serna, A. M. Somoza, and M. Ortuño, *Loop Models with Crossings*, Phys. Rev. B **87**, 184204 (2013).
- A. Nahum, J. T. Chalker, P. Serna, M. Ortuño and A. M. Somoza, *Phase Transitions in Three-Dimensional Loop Models and the* CP^{n-1} *Sigma Model*, Phys. Rev. B **88**, 134411 (2013).
- A. Nahum, J. T. Chalker, P. Serna, M. Ortuño and A. M. Somoza, *Length Distributions in Loop Soups*, Phys. Rev. Lett. **111**, 100601 (2013).
- In preparation: A. Nahum, J. T. Chalker, P. Serna, M. Ortuño and A. M. Somoza, *Deconfined Criticality in a Classical Loop Model*; P. Serna, M. Ortuño, A. M. Somoza, A. Nahum and J. T. Chalker, *Critical Behaviour in 3D Unoriented Loop Models*; A. Nahum, *RG Flows for* Θ*-Point Polymers*; A. Nahum, *Field Theories for Fully-Packed Loop Models and RVB States.*

Supervisor's Foreword

It is a pleasure to write this foreword for Adam Nahum's thesis. When he started the research, I had little idea how far it would go or how wide-ranging it would turn out to be.

Problems involving loops or random curves occupy an important place in theoretical physics. A good example is provided by the connection that P.-G. de Gennes established, between self-avoiding walks in polymer physics and the high-temperature expansion for the $O(n)$ model in the $n \rightarrow 0$ limit. Nevertheless, one might easily have thought 5 years ago that most of the interesting points had already been made. The thesis shows in multiple ways that this was not the case.

One of the appeals of the results described in the following pages comes from seeing current topics in a new light. An instance is the formulation of a particularly subtle phase transition in quantum magnets as a loop problem in classical statistical physics, with all the simplifying features of real, positive configurational weights replacing the phases and interference effects of quantum systems. Another is the connection between certain exotic Anderson localisation problems and the phase transitions discovered here in two-dimensional loop models.

A second attraction of this work comes from the understanding it gives of old results. For example, numerical studies of random curves in a variety of settings have found they have a fractal dimension close to two. The field-theoretic description developed here shows how this stems naturally from standard properties of Goldstone modes, with two as the exact value.

Research should sometimes lead to surprises. For me in the following it is to find an exact statement that can be made about a strongly correlated statistical system—involving the probability distribution of loop lengths, which turns out to be a non-trivial yet calculable function of an infinite number of degrees of freedom.

Computational work is frequently crucial in theoretical physics and collaborations can be key even in small-scale science. Adam was fortunate in having expert co-workers from the University of Murcia, whose simulations have provided tests and substantiation of many of the ideas here.

Looking ahead, classical loop models seem set to play an important part over the next few years in thinking about strongly correlated quantum phases. I hope some of the results in this thesis will be useful in that endeavour.

Oxford, April 2014 Prof. John Chalker

Abstract

Many critical systems are best discussed in the language of random geometry, in particular of random curves—which might be worldlines, line defects, Feynman graphs, polymers, etc. This thesis explores such formulations, aiming to extend the geometric language in places where it is underdeveloped: three-dimensional (3D) loop ensembles in the first half of the thesis and 2D loop models lying outside the standard paradigms in the second. Many of the problems studied here relate to sigma models on complex or real projective space (CP^{n-1} or RP^{n-1}), and this is also an investigation of critical behaviour in these theories.

We begin with a family of classical 3D loop models that show transitions between phases with and without infinite loops. We map them to the above field theories, with n fixed by the fugacity for loops. Treating n as continuously varying, we characterise both renormalisation group flows in the CP^{n-1} model and the universal statistical geometry of the loops (and generalise the latter conclusions to some other 3D ensembles). We argue that certain of the models exhibit the physics of deconfinement and emergent gauge fields, and connect them to 2D quantum antiferromagnets.

Next we tackle the fractal geometry of vortex lines in disordered media. These show geometrical phase transitions analogous to percolation transitions, but, we argue, in distinct universality classes. Vortex geometry has been studied numerically in many contexts, but field theory descriptions have been lacking. Via mappings to lattice gauge theory, we argue that replica limits or supersymmetric versions of the above theories are the required descriptions, and use them to classify critical properties.

2D loop models are well studied—but there is more to discover if we look beyond the familiar ones in which loop crossings are forbidden. We reveal new universality classes of continuous phase transitions, driven by the unbinding of $\mathbb{Z}_2$ point defects in the RP^{n-1} sigma model (with $n < 2$). We also characterise in detail the 'Goldstone' phase which appears very generically for loops with crossings. One of our models provides a close analogue for Anderson metal-insulator transitions, and a simpler context in which to study such poorly understood central-charge-zero critical points.

In the final chapter we address some long-standing questions concerning the polymer collapse transition (Θ point). We give a field theory treatment for the standard lattice model of a polymer with crossings in 2D, explaining previous numerical results. Surprisingly, the collapse transition in this model turns out to be governed by an infinite-order multicritical point, indicating that more generic models will show new universal behaviour.

Acknowledgments

Most importantly, I am immensely grateful to my supervisor John Chalker for sharing his great wisdom and insight. I have learned an enormous amount from working with him and drawn inspiration from his distinctive approach to physics. His generous and patient mentorship often went beyond the call of duty, and I thank him wholeheartedly.

John Cardy has been a great source of inspiration. I thank him for conversations about field theory and random geometry which have shaped my thinking on these subjects, and for his valuable encouragement in this work.

I thank Ilya Gruzberg for everything he taught me at the University of Chicago and his excellent guidance. I learned a great deal from him, and also from Eldad Bettelheim and Paul Wiegmann, and their enthusiasms helped shape my own.

My sincere thanks go to Andres Somoza, Pablo Serna and Miguel Ortuño for an enjoyable and educational collaboration. Without their simulations, much of this work would have been impossible.

I have benefited from discussions with many people regarding the work presented here. In addition to those above, I thank Fabien Alet, Szilard Farkas, Paul Fendley, Jesper Jacobsen, Ludovic Jaubert, Curt von Keyserlingk, Roderich Moessner, Thomas Prellberg, Kirill Shtengel, and particularly Fabian Essler (for robust exchanges!) and Andreas Ludwig.

I also warmly thank Joe Bhaseen, Fiona Burnell, Claudio Castelnovo, Ben Davison, Jerome Dubail, Tarun Grover, Dmitry Kovrizhin, Oleg Lunin, Luke Pagarani, Siddharth Parameswaran, Mark Rosin, Steve Simon and Arvind Subramaniam for their guidance. I am indebted to Adam Brown, who helped me find my feet in physics as an undergraduate and continued to provide sage advice, and to Roxana Mihet for helpful encouragement when I returned to Oxford. I thank my colleagues at the Rudolf Peierls Centre for making it a congenial place to work, and Simon Davenport for putting up with me as an officemate. In particular I thank Patrick Draper for his counsel in many things and for being an indispensable comrade over the years.

Lastly, it makes me very happy to thank Emmanuelle Debouverie for everything, including her wonderful support while I was writing this. And of course my family, Chloe, Fiona and Andrew Nahum, to whom I dedicate this.

Contents

Abbreviations

CPLC	Completely packed loop model with crossings
IPLC	Incompletely packed loop model with crossings
ISAT	Interacting self-avoiding trail
$NCCP^{n-1}$ model	Non-compact CP^{n-1} model
SAW	Self-avoiding walk
SLE	Schramm–Loewner evolution

Chapter 1
Introduction

1.1 Loop Models and Critical Phenomena

Critical systems can typically be regarded as fluctuating soups of fractal objects. These objects may be immediately visible in the spatial configurations of the system—for example domain walls in an Ising lattice magnet, or polymer chains in a dilute solution—or they may appear only after expressing the system's partition function in terms of new degrees of freedom. Examples of the second type include graphs in the high-temperature expansions of classical lattice models and worldlines of quantum particles in spacetime. Often the fractal objects are random walks or loops of some kind, but more generally they could be clusters, membranes, trees, nets, etc. Geometrical formulations are ubiquitous, and for this reason random geometry provides a very general language for critical phenomena.

Translating problems into this language may expose hidden relationships between them, because similar geometrical ensembles can often be reached from different starting points. For instance, the self-duality of the two-dimensional Ising model (relating the ordered and disordered phases) arises because the low and high-temperature expansions generate similar ensembles of loops, while the analogy between worldlines in two-plus-one dimensions and vortices in three leads to dualities between bosonic models. More surprising relationships abound too. For example, it turns out that a single universality class embraces both classical phase transitions for polymer chains and transport properties for quasiparticles in certain disordered superconductors [1–4].

Focusing on random geometry also leads to highly effective calculational schemes. These include numerical methods, such as the Wolff Monte Carlo algorithm (which updates spins in clusters rather than one at a time) or algorithms for quantum systems that use a worldline representation [5, 6]; analytical techniques, for example the Coulomb gas technique in two-dimensional statistical mechanics [7]; and even rigorous formalisms such as Schramm–Loewner evolution [8].

The language of random geometry, as well as being useful for answering questions about critical systems, is sometimes indispensable for talking about them at all.

A. Nahum, *Critical Phenomena in Loop Models*, Springer Theses,
DOI 10.1007/978-3-319-06407-9_1

Percolation is the paradigmatic example of this. Correlation functions of local degrees of freedom are trivial, and all the interest is in 'geometrical correlation functions' that encode nonlocal information, like the probability of two sites lying in the same cluster. Polymer conformations yield another problem for which a geometrical language is clearly inevitable. Even for the most familiar critical systems, such as the Ising model, geometrical observables are part of the full structure of the critical theory; we may think of the usual field theories for the Ising model (like scalar ϕ^4 theory) as capturing only a sector of a larger and richer structure.

Extending the field theory treatment of critical systems to geometric observables leads us to some of the most exciting problems in critical phenomena. At the same time, it takes us into a lawless domain where familiar guiding principles (like the Mermin Wagner theorem or Zamoldchikov's c-theorem in two dimensions [9]) may no longer apply, and where hidden symmetries may complicate the assignment of systems to universality classes. Geometrical phase transitions may be 'non-Landau-Ginzburg' in the sense that the observables of interest are not always captured by fluctuations of a local order parameter. In such cases we can make a great deal of progress by conjuring an order parameter into existence,[1] but we should be ready for surprises.

This thesis focusses on critical phenomena that have a connection to loop models. These are simply statistical ensembles of loops on the lattice. In line with what was said above, they are not only a special class of statistical mechanics problems but also a useful language for critical phenomena more generally. Within this framework we will be able to make contact with questions arising from quantum magnetism, classical disordered systems, polymer physics and (more obliquely) disordered fermions, among other areas. The connection with quantum magnets is that worldlines of spinons form loop ensembles in space-time, and that with classical disordered systems is through the geometry of their topological line defects. The connection with polymer physics is of course that long linear polymers, or idealised models for them, yield scale-invariant random curve ensembles. Disordered fermions will not be treated explicitly, but have close connections with critical phenomena in the 'replica' sigma models we discuss.

The problems considered here are also tied together by the field theories describing them, which are mostly related to sigma models on real or complex projective space (RP^{n-1} and CP^{n-1}). This thesis is also an investigation of critical behaviour in this class of theories, which is important in diverse areas of theoretical physics.

We begin this introduction by reviewing a class of two-dimensional loop models that has been very influential for the geometrical approach to critical phenomena (Sects. 1.2–1.4). We briefly discuss polymers, another driver of this approach (Sect. 1.5), and then a class of quantum problems that have a relation to geometrical ensembles in space rather than in space-time (Sect. 1.6). The field theories

[1] For example, for percolation we can employ a well-known correspondence with the q state Potts model in the limit $q \to 1$ [10].

we will encounter are mentioned only briefly (Sect. 1.7) since we will give further pedagogical material about the CP^{n-1} model (and its cousin the $NCCP^{n-1}$ model) in Chaps. 2 and 3. In Sect. 1.8 we review some connections between disordered systems and random geometry. Finally, Sect. 1.9 is an overview of this thesis.

1.2 Loops on the Honeycomb Lattice

The '$O(n)$' loop model on the honeycomb lattice is an important paradigm for two-dimensional (2D) statistical mechanics, because it gives a unifying picture for critical behaviour in a remarkable range of two or one-plus-one-dimensional systems.

It is a useful point of departure for us (even if the next three chapters will be about 3D problems) as it illustrates basic ideas about loop models and their relation to spin models, and also some of the subtleties that can arise in extracting field theories for geometrical problems. In particular it brings us to the issue of crossings (intersections) in 2D loop models [9, 11, 12] which will be crucial in Chaps. 5 and 6, and the underappreciated fact that the underlying global symmetry of the honeycomb lattice model is really $SU(n)$ rather than $O(n)$.

The following is a brief overview—for introductions to some of the exact techniques available for non-crossing loops in 2D, see Refs. [7, 13–16].

Figure 1.1 (left) shows part of a configuration of non-intersecting loops on the honeycomb lattice. The partition function for the loop model is a sum over such configurations, weighted by a fugacity x for the total loop length and a fugacity n for the number $|\mathcal{C}|$ of loops in a configuration $\mathcal{C}$:

$$Z = \sum_{\mathcal{C}} x^{\text{length}} n^{|\mathcal{C}|}. \tag{1.1}$$

This model is usually referred to as the $O(n)$ loop model because (when n is an integer) it may be mapped to a particular variant of the $O(n)$ magnet on the honeycomb lattice [17, 18]. This name is misleading in some respects, as we will see [9].

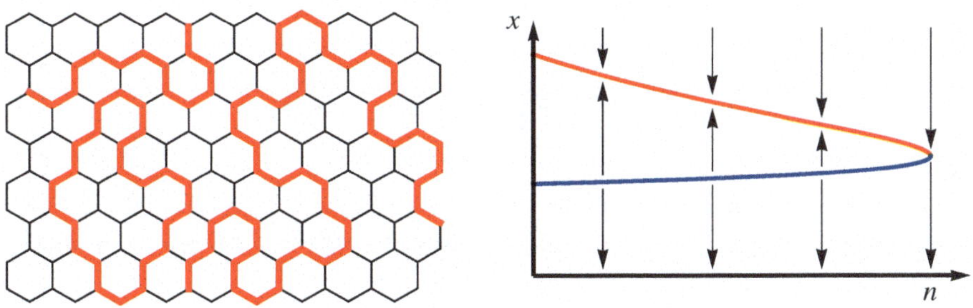

Fig. 1.1 *Left* loops in the honeycomb lattice $O(n)$ loop model. *Right* schematic RG flows in this model for $0 \leq n \leq 2$. The dense (*upper*) and dilute (*lower*) fixed points coalesce at $n = 2$

A schematic RG flow diagram is shown in Fig. 1.1 (right). For small length fugacity, loops are short. At an n-dependent critical fugacity x_c (blue line in figure) the renormalised line tension vanishes and the loops become scale-invariant random fractals that are characterised by universal exponents. The most basic one is the fractal dimension d_f, which governs how the typical linear size R of a loop scales with its length ℓ:

$$\ell \sim R^{d_f}. \tag{1.2}$$

The point x_c is referred to as the 'dilute' critical point. For $x > x_c$, loops are again critical but with distinct universal behaviour; this is referred to as the 'dense' phase. Note that n does not flow under RG, and the exponents are n-dependent.

The $O(n)$ magnet [17, 18] that maps to (1.1) is obtained by placing n-component spins $\mathbf{S}_i$ (with $\mathbf{S}_i^2 = n$) at the sites i of the lattice and choosing for their partition function[2]:

$$Z = \mathrm{Tr} \prod_{\langle ij \rangle} \left(1 + x\, \mathbf{S}_i . \mathbf{S}_j \right). \tag{1.3}$$

In general, the above partition function differs from the more standard

$$Z = \mathrm{Tr} \exp \left(J \sum_{\langle ij \rangle} \mathbf{S}_i . \mathbf{S}_j \right), \tag{1.4}$$

though the two forms coincide at $n = 1$ (when the S_i become Ising spins) with $x = \tanh J$. But since (1.3) retains the $O(n)$ symmetry and the ferromagnetic nature of (1.4), we might surmise that the universal critical properties will be identical. We will return to the question of whether this is true below.

To see the relation to the loop model, expand out the product over links $\langle ij \rangle$ in (1.3) and represent the terms graphically by the rule that we colour each link for which we choose '$x\, \mathbf{S}_i . \mathbf{S}_j$' rather than '1'. Integrating over the spins, only the terms with an even power of $\mathbf{S}_i$ at each node i survive. This forces the links in the graphical expansion to form configurations of nonintersecting loops. Letting $\mathcal{L}$ denote a loop in a configuration $\mathcal{C}$, we have

$$Z = \sum_{\mathcal{C}} x^{\text{length}} \prod_{\mathcal{L} \in \mathcal{C}} \mathrm{Tr}\, (\mathbf{S}_1 . \mathbf{S}_2) \ldots (\mathbf{S}_\ell . \mathbf{S}_1). \tag{1.5}$$

(We use a loose notation in which the sites on a given loop $\mathcal{L}$ are labelled $1, \ldots, \ell$.) Next, the integral over the spins is performed using $\mathrm{Tr}\, S_i^\alpha S_i^\beta = \delta^{\alpha\beta}$, $\mathrm{Tr}\, 1 = 1$, giving

[2] Throughout this thesis, the trace 'Tr' will denote an integral over degrees of freedom, here the fixed-length vectors $\mathbf{S}_i$, while 'tr' will denote a matrix trace.

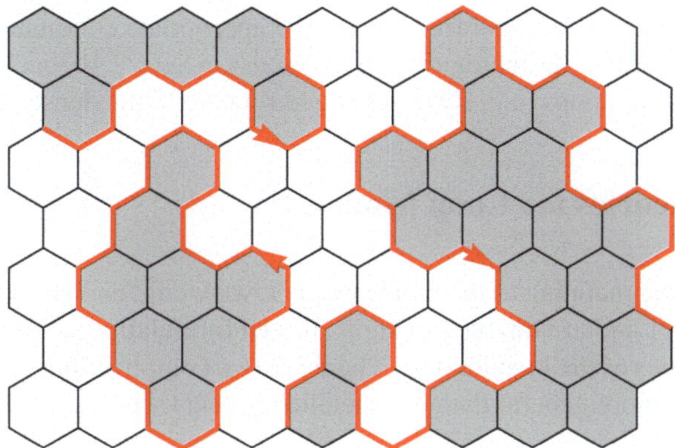

Fig. 1.2 At $n = 1$, the loops may be viewed as cluster boundaries in the Ising model, or in uncorrelated percolation if $x = 1$. For general n, regarding the loops as cluster boundaries also gives a prescription for fixing their relative orientations

$$\mathrm{Tr}\,(\mathbf{S}_1.\mathbf{S}_2)\ldots(\mathbf{S}_\ell.\mathbf{S}_1) = \mathrm{Tr}\,(S_1^{\alpha_1} S_2^{\alpha_1})(S_2^{\alpha_2} S_3^{\alpha_2})\ldots(S_\ell^{\alpha_l} S_1^{\alpha_l}) = \delta^{\alpha_1\alpha_2}\delta^{\alpha_2\alpha_3}\ldots\delta^{\alpha_l\alpha_1}. \tag{1.6}$$

Repeated indices are summed throughout. The delta functions leave a single 'colour' index α, summed over $\alpha = 1, \ldots, n$, for each loop. We may either think of this as giving an ensemble of coloured loops, or we may sum over the colours to yield a statistical weight n per loop as in (1.1).

In the case $n = 1$, the graphical representation (1.1) leads us directly to the duality between the high and low temperature phases of the 2D Ising model. We see this by reinterpreting the loops as domain walls for a dual Ising model. The dual spins, represented by grey and white hexagons in Fig. 1.2, live on a triangular lattice and are coupled with strength J^*, given by $x = \exp(-2J^*) = \tanh J$.

The original and the dual spins are critical at the dilute critical point x_c. In the dense phase ($x > x_c$), the original spins S_i are ordered and the dual spins disordered. Neither are critical, although the loops are. In fact the $n = 1$ dense phase describes percolation cluster boundaries: this is clear from the fact that at $x = 1$ ($J^* = 0$) the dual spins yield (critical) uncorrelated site percolation on the triangular lattice. The fractal dimension of a percolation cluster boundary is $7/4$ while that of a critical Ising domain wall is $11/8$; in general, the dense loops are more compact than the dilute ones (see Sect. 1.4).

So far, we have ignored the fact that the Boltzmann weight for the Ising spins S_i can become negative if x exceeds one in Eq. 1.3. This does not lead to a pathology in the loop model. This regime lies in the dense phase, with the exception of the limiting value $x = \infty$, for which every site is visited by a loop.[3] This full-packing

[3] For $n = 1$, $x \to \infty$ we have $J^* \to -\infty$, so in the dual language this is the triangular lattice Ising antiferromagnet at zero temperature, which is critical as a result of frustration.

constraint leads to an additional free field in the appropriate continuum description, by a mechanism whose 3D analogue we will discuss in Sect. 2.10, but leaves many of the properties of the loops (e.g. d_f) unchanged relative to the dense phase [19, 20].

1.3 Field Theories for Loop Models

On the basis of the mapping to the lattice magnet, we would naively expect the $O(n)$ field theory (in a Landau-Ginzburg or sigma model formulation) to describe the long distance properties of the loop models. There are two complications, which illustrate features that are more general than the specific model (1.1).

1.3.1 Replica-Like Limits

We saw above that while the loops are critical in the dense phase for $n = 1$, the Ising spins are not. Ising correlation functions therefore do not suffice to capture the fractal geometry of the loops. Similar problems arise in the dilute phase and at other integer values of n; the inadequacy of the standard $O(n)$ model is even clearer if we wish to consider the loop model at noninteger n or in the limit $n = 0$, where it yields important models for polymers (Sect. 1.5). Thus we would like to expand the range of correlation functions that we can express.

A solution is to formally treat n as an integer of arbitrary size and then analytically continue to the desired value. The correlation functions of local operators, when analytically continued, have a meaning in the loop model. For example, the two-point function of the local operator $S^1 S^2$ (an operator which does not exist if we set $n = 1$ or $n = 0$ from the outset) is proportional to probability that two sites lie on the same loop, and its scaling dimension is related to d_f, while the two-point function of S^1 gives a sum of configurations in which the two points are joined by an open curve.

This kind of limit was introduced by de Gennes for the description of polymer statistics [21]. A similar idea from around the same time is analytical continuation in the q of the q-state Potts model, which yields relations to percolation and spanning trees [10]. These limits are closely related to the replica trick in disordered systems (Sect. 1.8) and for want of a better name we will call them replica-like limits. Quite apart from their appearance in the context of random geometry, analytic continuations in parameters like n and q may give us a useful handle on conventional criticality, in the same way that analytic continuation in the number of spatial dimensions does [22] (see Sect. 2.8.2).

An alternative way to extend the spectrum of operators in the field theory is using supersymmetry (SUSY) [23, 24]. We defer discussion of this until Chap. 2. Both replica-like limits and SUSY are standard tools for geometrical critical phenomena.

1.3.2 Hidden Symmetry and the Appearance of CP^{n-1}

If our starting point is the loop model (1.1), the utility of a spin formulation like (1.3) is that it reveals symmetries that are hidden in the geometrical formulation. But how do we know there is not more symmetry hiding? This is a general issue for geometrical problems (polymers, etc.). It is always possible to map the configuration sum to a partition function for local (e.g. spin) degrees of freedom in more than one way, and some formulations may not expose all the symmetries or conservation laws. As a result we can be misled about the continuum description.

In particular it is not always immediately obvious whether a constraint on the allowed geometrical configurations amounts to an irrelevant microscopic detail or whether it changes the continuum description. Here we are interested in the fact that the loops in the honeycomb lattice model cannot cross, but let us briefly mention that the full-packing constraint ($x = \infty$) in the honeycomb model is another example (Sect. 1.2). This creates an extra conservation law, changing the continuum limit.[4] Other examples of hidden symmetry are associated not with constraints on geometrical configurations but with fine-tuning of their weights—in Chap. 6 we discuss an example that has led to confusion over the field theories for certain polymer models.

In the spin formulation, the no-crossing constraint in the honeycomb loop model relies on the specific truncated Boltzmann weight (Eq. 1.3): it is lost if higher powers of $(\mathbf{S}_i.\mathbf{S}_j)$ are included or if a lattice of higher coordination number is used. It does lead to higher symmetry and in some cases a different continuum description. The enhanced symmetry for non-crossing loops or worldlines has been noted in various contexts, but the implications for the universal behaviour of the loop models has been understood only relatively recently by Read, Saleur and collaborators [9, 11, 25–27].

Let us give a heuristic sketch of why a different continuum description appears.[5] Note first of all that an alternative way to think of the lattice magnet (1.3) is as a regularisation of a field theory for a quantum problem in one spatial dimension and imaginary time. In this interpretation, the loops are simply worldlines of the n 'colours' of quantum particles associated with the fields S^α.

When loops don't cross, there is a simple prescription for fixing their relative orientations: if we regard them as cluster boundaries, they can be oriented by the rule that grey clusters are encircled anticlockwise (Fig. 1.2). In the quantum language, the appearance of oriented worldlines means that we can associate a $U(1)$ charge with the particles. We therefore promote the real field $\mathbf{S} = (S^1, \ldots, S^n)$ to a complex one $\mathbf{z} = (z^1, \ldots, z^n)$.

---

[4] Using the bipartiteness of the honeycomb lattice, one can define a vector field J on the links which is divergence free when loops are fully packed. In a continuum notation, $\nabla_\mu J_\mu = 0$. Crudely, J is related by $J_\mu = \epsilon_{\mu\nu}\nabla_\nu\varphi$ to the additional free field φ that appears in the continuum description when $x = \infty$.

[5] For a simple explicit derivation in the honeycomb model see Sect. 4.5; a similar approach is in Ref. [27]. For the completely-packed models (Sect. 1.4), the continuum description may be seen by a mapping to a spin chain [11, 25] or an appropriate lattice field theory (Chaps. 2 and 3).

By introducing orientations, we give up the strict locality of the interactions between particles. Adjacent loops are forced to be oppositely oriented, meaning in the quantum language that adjacent particles must be oppositely charged, regardless of their separation. In the continuum, this is associated with a long-range interaction mediated by a $U(1)$ gauge field with a θ term [11, 27, 28]:

$$\mathcal{L} = \frac{1}{2g} |(\nabla - iA)\mathbf{z}|^2 + \frac{i\theta}{2\pi}\epsilon_{\mu\nu}\nabla_\mu A_\nu, \quad |\mathbf{z}|^2 = 1. \tag{1.7}$$

The coefficient θ is equal to π. With this choice, the gauge field yields an interaction with the desired properties. In fact in the quantum language, the θ term simply provides a background electric field, and the alternation of positively and negatively charged particles has a simple energetic explanation (given for example in [29]).

This field theory is one of the formulations of the CP^{n-1} nonlinear sigma model. We will return to this class of field theories repeatedly in this thesis, in various formulations and in two and three dimensions. For present purposes, the important point is the unexpected appearance of an $SU(n)$ symmetry (global rotations of $\mathbf{z}$) as a result of the no-crossing constraint. Introducing crossings in general breaks this symmetry [9, 12, 25] (see Chap. 5 for more details). The question is thus whether this symmetry-breaking is RG relevant or irrelevant.

It turns out that it is irrelevant in the dilute phase, but relevant in the dense phase [9]. We therefore expect that a generic $O(n)$ loop model, including crossings, will show the same *dilute* critical behaviour as the honeycomb model[6] but that the *dense* phase will be replaced with something else [9]. Loops with crossings are a fertile area for new types of critical behaviour [9, 12] which we will explore in detail in Chaps. 5 and 6.

1.4 More 2D Background

The fixed points in Fig. 1.1, both dense and dilute, form a single line along which the critical exponents vary continuously. A wide variety of systems (in addition to those encountered above) can be associated with points on these lines. We briefly survey some of them and the alternative continuum descriptions that exist.[7]

[6] There is a caveat at $n = 2$—a more generic lattice XY model would show continuously varying critical exponents: this doesn't happen in (1.3), as the RG flow diagram (Fig. 1.1) makes clear. This is a result of the hidden $SU(2)$ symmetry, which constrains the model to the flow line leading to the KT fixed point. Even at $n \neq 2$ a model with crossings may show additional corrections to scaling associated with irrelevant symmetry-breaking perturbations.

[7] We discuss here only the case $n \geq 0$, but the loop model may also be defined for $n < 0$. For example, $n = -2$ is related to the so-called loop-erased random walk and to paths on spanning trees.

1.4.1 Potts Model and Completely Packed Loops

The 2D Potts models may be mapped to loop models via the Fortuin Kasteleyn representation [10]. The results look at first sight rather different from the honeycomb lattice model above: they are 'completely packed' models on the square lattice, in which every link is visited by a loop (see Chap. 2). However, they share the property that the loops do not intersect, and turn out to be another realisation of the dense phase. Thus the critical q-state Potts model may be related to the upper branch in Fig. 1.1: the relationship is $n = \sqrt{q}$. (So the critical Ising model is related to the point $n = \sqrt{2}$ on the dense branch, as well as the point $n = 1$ on the dilute branch—this is an example of the SLE duality discussed below.)

1.4.2 Spin Chains

In an anisotropic limit, the above completely-packed loop models may be mapped to one-dimensional quantum antiferromagnets with $SU(n)$ symmetry, so for example the critical point at $n = 2$ is in the universality class of the spin-$1/2$ antiferromagnetic Heisenberg chain. We will review this idea in more detail in Chap. 3. The spin chains also have supersymmetric generalisations [1, 11, 25].

1.4.3 SLE

The one-parameter family of universality classes seen in Fig. 1.1 also appears in Schramm–Loewner Evolution, or SLE [8]. This is a mathematically rigorous approach to two-dimensional critical phenomena which is formulated directly in the continuum, and which focusses on non-crossing random curves. The statistics of these curves are encoded in a stochastic differential equation governing the *growth* of an open one, starting at a boundary, as a function of a fictitious time (see Refs. [13, 16] for reviews).

Under certain assumptions, including the conformal invariance of the measure on the curves, the only free parameter in this stochastic equation is the strength of the noise, κ, and this determines the universal properties of the curve; for example its fractal dimension is $d_f = 1 + \kappa/8$. The correspondence with the critical points in the $O(n)$ model is given by $n = -2\cos(4\pi/\kappa)$, with $4 < \kappa \leq 8$ for the dense phase and $2 \leq \kappa \leq 4$ for the dilute phase. The central charge of the $O(n)$ loop model may also be expressed in terms of κ,

$$c = 13 - \frac{3}{2}\left(\kappa + \frac{16}{\kappa}\right). \tag{1.8}$$

The ‘dual’ values κ and $\tilde{\kappa} = 16/\kappa$ have the same central charge, and there is a geometrical relationship between them. One of the points lies in the dense, and the other in the dilute regime. In the former (but not in the latter) the SLE curve continually touches itself.[8] It is then possible to define an outer boundary (hull) for the curve which is distinct from the curve itself, and the statistics of the hull are those of the dilute curve at the dual value of κ. At $c = 0$, this duality is between percolation cluster boundaries ($\kappa = 6$) and self-avoiding walks ($\kappa = 8/3$).

One of the remarkable things about SLE is the very constrained classification it provides of universal behaviour in two dimensional (non-intersecting) loop models. This should be taken with a grain of salt, since (1) a given lattice model may fail to obey the postulates of SLE, and (2) SLE does not give access to all the universal properties of the model.[9] However it does suggest that in order to see genuinely new behaviour for 2D loop models, rather than just different lattice regularisations of the same set of fixed points, it is a good idea to relax the non-crossing constraint.

1.4.4 Coulomb Gas

Finally we mention the Coulomb gas approach to loop models without crossings. When $n = 2$, we can reproduce the loop fugacity in Eq. 1.1 by summing over the two possible orientations for each loop. This soup of randomly oriented loops can then be mapped to a height model: the height h is defined on the hexagons, and takes integer values which jump by one every time a loop is crossed, with the sign of the jump determined by the loop's orientation. By construction there is a symmetry $h \to h \pm 1$. The continuum description can be taken to be simply to be the free Lagrangian for h, supplemented by the symmetry-allowed term $\cos 2\pi h$. This is the well-known dual (solid-on-solid) representation of the 2D XY model. As a result of the hidden $SU(2)$ symmetry, the couplings are such that it flows to the KT fixed point for all $x \geq x_c$.

Surprisingly, this approach has an elegant extension to $n \neq 2$. Random orientations are again assigned to the loops, but now a phase $e^{\pm i\chi}$ is associated with every node visited by a loop—the sign is determined by whether the loop turns clockwise or anticlockwise (i.e. by $\pm\pi/3$ radians) at the node. For each loop, the total turning angle is $\pm 2\pi$, which means that the total number of clockwise turns minus the total number of anticlockwise turns is ± 6. Summing over the orientations thus generates a weight $e^{6i\chi} + e^{-6i\chi}$; we choose χ so that this equals the desired weight n. In the continuum, h is described by a free field theory modified with screening operators and boundary terms; calculations are still possible. See Refs. [7, 13–15] for more details, and Refs. [30–32] for some extensions to other loop models.

[8] This is connected with the RG relevance of crossings in the dense phase.

[9] For example, the loops in the dense phase and in the fully-packed limit $x = \infty$ of the honeycomb lattice model correspond to the same κ, but appropriate correlation functions will reveal the existence of an extra massless mode at $x = \infty$.

Unfortunately an extension of the above approach is not known for the models with crossings discussed in Chap. 5. (The χ trick relies crucially on the fact that the loops do not cross themselves—if they do, a loop can turn through any multiple of 2π radians.) For models *without* crossings, the Coulomb gas description and the CP^{n-1} description (1.7) are complementary: while the former allows exact calculations of scaling dimensions etc., the latter makes the symmetries of the problem manifest and allows us to consider perturbations (like the introduction of crossings) that the Coulomb gas is not set up to deal with.

Another important set of applications of dilute and dense loops is to polymers and related problems, which we now turn to.

1.5 Polymers

Many fascinating questions have arisen from the statistics of polymer conformations, not all of which are answered. Geometric critical exponents for polymers of course feed into more 'physical' observables: for example the viscosity of a polymer solution will depend on the typical size of a conformation, and thus its variation with polymer length will involve the fractal dimension [33]. More abstractly, polymer models have spurred the development of field theoretic tools for critical systems, including replica-like limits [21, 34] and various kinds of supersymmetry [23, 24, 35], and they continue to be important in the development of logarithmic conformal field theory [36–39].

The honeycomb lattice loop model yields models for a linear polymer at three points on the critical line. Two of these occur at $n = 0$: when the loop fugacity vanishes, the only loops that survive in the graphical expansion are those that are 'put there' by operator insertions, so this limit can be used to describe the statistics of a single loop or strand.

The dilute critical point at $n = 0$ describes the universality class of the self-avoiding walk (SAW), or the swollen phase of a long polymer with excluded volume interaction.

The dense phase on the other hand describes a single loop or strand that visits a finite fraction of the sites of the lattice. This universality class is also related to the conformations of a polymer in the collapsed phase caused by sufficiently strong self-attraction. (The 'dense polymer' is also related to various other problems, including Hamiltonian and Eulerian paths,[10] uniformly weighted spanning trees,[11] and even variants of the travelling salesman problem [40].)

At first sight this exhausts the applications of the honeycomb lattice model to the statistics of a *single* polymer chain, since when $n \neq 0$ the configurations are soups of many loops. Nevertheless when $n = 1$ and $x = 1$—i.e. for percolation cluster

[10] Paths which visit all the sites or all the links of a lattice, respectively.

[11] Loop-free clusters that visit all the sites of a lattice. The boundary of such a tree is related to the dense polymer.

boundaries—there is a simple mapping to a model for a single self-avoiding loop with a self-attraction of a certain type [4, 41]. This describes a polymer at the so-called Θ point, which is the boundary between the SAW phase and the collapsed phase, and exact exponents for this problem can be obtained from the Coulomb gas [4].

De Gennes pointed out that the Feynman diagrams for the linear polymer with an excluded volume interaction are equivalent to those for the $O(n)$ model, once the limit $n \to 0$ is used to suppress loops [21]. The above description of the SAW is expected to be in the same universality class as the critical $O(n)$ model, since crossings are irrelevant in the dilute phase, and this model is an elegant realisation of de Gennes' correspondence. The universal properties of the dense polymer on the other hand are sensitive to the presence or absence of crossings [9].

Behaviour at the Θ point is more subtle, and different models can show different universality classes of collapse transition. In Chap. 6 we will consider the universal behaviour of models for polymer collapse in which self-avoidance is not strictly enforced, so that the polymer is allowed to cross itself.

Other interesting directions include melts or networks of polymers, polymers in random environments [42], and polymers with intrinsic randomness. A problem with some unusual features is that of a randomly branched polymer. This turns out to have a powerful supersymmetry which leads to a dimensional reduction to a problem in two fewer spatial dimensions. Thus exact exponents are available in the 'physical' case $D = 3$ [35, 43, 44].

1.6 Quantum Random Geometry

So far we have mostly discussed classical models. Quantum systems may also be usefully viewed in terms of extended degrees of freedom, in either space or spacetime.

Ensembles of worldlines in spacetime are widely used for Monte Carlo simulations [45]. Working with extended objects allows for more efficient sampling, either because nonlocal updates are possible (like changing a 'colour' index of a worldline) or because degrees of freedom can be summed over analytically (like summing over colours to give a fugacity for worldlines).

Mappings of quantum problems in D spatial dimensions to classical loop ensembles in $D + 1$ are examples of the usual quantum-classical correspondence in statistical mechanics. But D dimensional quantum many-body systems may also permit mappings to geometric ensembles in D dimensions. These are of particular interest in the context of topological phases [46–55].

One of the simplest examples is for a magnet with spin-$1/2$ degrees of freedom on each link of a lattice, say the honeycomb [48–52]. For an appropriately engineered Hamiltonian, the ground state is a sum over configurations (denoted $\mathcal{C}$) in which the up spins form nonintersecting loops: $|\Psi\rangle \propto \sum_{\mathcal{C}} \Psi(\mathcal{C})\,|\mathcal{C}\rangle$. Since $\langle\mathcal{C}|\,\mathcal{C}'\rangle = \delta_{\mathcal{C}\mathcal{C}'}$, expectation values of operators that are diagonal in our chosen basis reduce to expectation values in a classical loop ensemble with partition function $Z = \sum_{\mathcal{C}} |\Psi(\mathcal{C})|^2$.

For the uniform superposition of loop configurations, $\Psi(\mathcal{C}) = 1$, we obtain the loop model discussed above at $x = 1$, $n = 1$ (percolation). This is the ground state structure of the toric code Hamiltonian, which realises a gapped (Abelian) topologically ordered state [48–52]. The presence of a gap relies on to the triviality of local correlation functions in percolation, while the presence of topological order relies on the vanishing line tension of the loops: for example, ground state degeneracy on the torus is associated with the possibility of loops which wind around the cycles, and deconfined anyons can be associated with endpoints of strands.

Various generalisations of this idea have been studied, including larger loop fugacities [49–52], i.e. $\Psi(\mathcal{C}) = \sqrt{n}^{\text{no. loops}}$, one of the aims being to find lattice models for gapped phases with non-Abelian topological order. In fact increasing the loop fugacity leads not to gapped topologically ordered states but to critical states, which are in themselves interesting. One way to pursue gapped states is to generalise from loops to branching nets [53–55]. Note that while states like $|\Psi\rangle$ above rely on fine-tuned Hamiltonians for their emergence, if the state is gapped then it should capture the generic properties of a phase. An interesting feature of these problems, whether gapless [49–52] or topologically ordered, is the way in which the random geometry of the classical ensemble feeds into the low energy properties.

In the above example, we had $\langle \mathcal{C} | \mathcal{C}' \rangle = \delta_{\mathcal{C}\mathcal{C}'}$. Resonating valence bond wavefunctions for spin-$1/2$ magnets are a case in which a nontrivial inner product plays a role (see also Ref. [54, 55]). Let $\widetilde{\mathcal{C}}$ denote a nearest-neighbour dimer covering of a given lattice, and let $|\widetilde{\mathcal{C}}\rangle$ be a state in which spins connected by dimers form singlets. Then consider the uniform superposition $|\text{RVB}\rangle \propto \sum_{\widetilde{\mathcal{C}}} |\widetilde{\mathcal{C}}\rangle$. Expectation values in this state do not directly reduce to those in a classical dimer model, because $\langle \widetilde{\mathcal{C}} | \widetilde{\mathcal{C}}' \rangle \neq \delta_{\widetilde{\mathcal{C}}\widetilde{\mathcal{C}}'}$. Instead, the associated classical partition function involves a sum over pairs of configurations $(\widetilde{\mathcal{C}}, \widetilde{\mathcal{C}}')$, weighted by $\langle \widetilde{\mathcal{C}} | \widetilde{\mathcal{C}}' \rangle$. Since superposing any two dimer configurations generates a configuration of loops (including loops of length two, i.e. dimers), we can reinterpret the sum on $(\widetilde{\mathcal{C}}, \widetilde{\mathcal{C}}')$ as the configuration sum for a loop model, and the Boltzmann weight turns out to be simple (for a bipartite lattice, with an appropriate sign convention for the singlets, loops have a fugacity of four and dimers have a fugacity of two). For studies of this problem in two and three dimensions see Refs. [56, 57]. We will return to it fleetingly in Sect. 2.11.

1.7 $\mathbf{CP}^{n-1}$ and $\mathbf{RP}^{n-1}$ Models

Degrees of freedom on real or complex projective space (RP^{n-1} or CP^{n-1}) play an important role in the following chapters. We have glimpsed the CP^{n-1} sigma model in Sect. 1.3.2; here we simply define real and complex projective space and describe some previous appearances of related field theories, leaving more detailed discussions until later chapters.

Complex projective space, CP^{n-1}, is the manifold of complex n-component unit vectors $\mathbf{z}$, subject to the identification of those differing by a phase:

$$\mathbf{z} \sim e^{i\phi}\mathbf{z}. \tag{1.9}$$

In some cases it will be useful to have a non-redundant parameterisation of CP^{n-1}, for example in order to consider soft-spin Lagrangians. This is provided for general n by the traceless Hermitian matrix

$$Q^{\alpha\beta} = z^{\alpha}\bar{z}^{\beta} - \frac{1}{n}\delta^{\alpha\beta}. \tag{1.10}$$

At the special value $n = 2$, the manifold CP^1 is simply the sphere, and we may instead form a three-component unit vector $S = (S^1, S^2, S^3)$ using the Pauli matrices:

$$S^i = \mathbf{z}^{\dagger}\sigma^i\mathbf{z} = \mathrm{tr}\,\sigma^i Q, \quad S^2 = 1. \tag{1.11}$$

The local 'spin' degree of freedom in the field theory (1.7) lives on CP^{n-1}, as a result of the gauge redundancy, and in fact this field theory can be simply mapped to a nonlinear sigma model (without a gauge field) for Q. At $n = 2$, this sigma model is simply the $O(3)$ sigma model. See Sect. 3.2 for more background.

CP^{n-1} models appear in many contexts and in various formulations. The 2D sigma model has been studied as a toy model for instanton effects in 4D gauge theories, making use of the fact that it has topological skyrmion textures for all n and is solvable in the large n limit via the saddle-point method [58, 59]. In condensed matter, the same field theory provides a description of quantum spin chains with $SU(n)$ symmetry, generalising the $O(3)$ sigma model for the antiferromagnetic $SU(2)$ Heisenberg chain [27]. In the $n \to 1$ limit, or in a SUSY formulation, the model also has applications to Anderson localisation (see Sect. 1.8).

We have mentioned the model's relevance to 2D loops above. In this context, sigma models on supermanifolds that generalise CP^{n-1}, denoted $\mathrm{CP}^{n+k-1|k}$, are also important [11, 25, 60]. Roughly speaking, we obtain $\mathrm{CP}^{n+k-1|k}$ by promoting $\mathbf{z}$ in (1.9) to a supervector with $n + k$ commuting (bosonic) and k anticommuting (fermionic) components (Q becomes a supermatrix). A cancellation between bosonic and fermionic loops means that the $\mathrm{CP}^{n+k-1|k}$ model may be mapped to the loop model with fugacity n independently of k; increasing k gives a richer spectrum of operators and is an alternative to the replica trick (these field theories are nonunitary).

In three dimensions we must distinguish between the CP^{n-1} model and the related 'noncompact' CP^{n-1} model, or NCCP^{n-1} [61–63], which will be described in Chap. 3. Both have applications to quantum magnetism in two dimensions, and the latter also describes the statistical mechanics of classical dimers in three dimensions [64–67]. These field theories are relevant to a large class of 3D loop models (Chap. 2), and the $n \to 1$ limit describes the statistics of topological defects in disordered systems (Chap. 4).

Real projective space, RP^{n-1}, is defined analogously to CP^{n-1}: we take an n-component real unit vector **S** and identify **S** with −**S**. The RP^2 sigma model ($n = 3$) arises in the context of nematic liquid crystals, and the well-studied Lebwohl Lasher model [68] constitutes a lattice regularisation of it. The case $n = 2$ gives simply an XY degree of freedom (RP^1 is the circle). We will show in Chap. 5 that the replica regime $n < 2$ describes interesting critical behaviour in 2D loop models.

The replica/SUSY sigma models arising from loop models indicate that the latter have much in common with disordered fermion systems. Another more direct connection of loops with disordered systems arises from linelike degrees of freedom in classical systems with quenched randomness. To clarify these connections we now recall some basic facts about disordered systems.

1.8 Disordered Systems

Let us begin with a crude distinction between three types of systems in which quenched disorder affects the critical properties.

Firstly, there are those where universal properties are directly determined by the random geometry of the disorder realisation. The well-known examples of this all reduce to the universality class of percolation. For instance, the geometry of clusters of magnetic sites determines the presence or absence of long range order in a dilute ferromagnet at zero temperature [69]. Another archetypal case is the quantum Hall transition in the semiclassical limit (where the magnetic length is much smaller than the typical scale for variations of the random potential). In this limit, eigenfunctions are confined to level lines of the random potential, which are continuum analogues of percolation cluster boundaries.

While percolation is the best-known universality class for geometrical transitions in disordered systems, one of the aims of Chap. 4 will be to show that it is not the only one. In three dimensions there is a richer variety of possibilities, associated with percolation-like transitions for line defects [28].

The second class contains idealised problems for which interactions may be neglected. Perhaps the most important are Anderson transitions for free fermions in disordered environments. These systems may be modelled by free field theories with random mass terms, gauge fields etc.

Finally, there are problems in which both interactions and disorder must be taken into account, but where the role of interactions is more complex than in the dilute magnet mentioned above. Spin glasses provide paradigmatically difficult examples, but there are wealth of classical and quantum problems involving random interactions, random fields, etc.

In reality the above distinctions are of course too simplistic, since the three classes bleed into one another. In two dimensions, the distinction between free and interacting models may be a matter of representation: for example random-bond Ising models and disordered dimer models can be mapped to free fermions and so become Anderson localisation problems. The idea of the geometry of the disorder configuration 'directly' determining universal properties is also vague—for example, do randomly diluted resistor networks fall into this class? There are also surprising connections between problems from different classes (one is mentioned below).

Nevertheless, the above classification is useful to locate geometrical problems in relation to disordered systems. In addition to the connection that exists by definition for the first class of problems, problems of the second type are closely related to the loop models that we will discuss, and they provide one of the motivations for wanting to understand some of the kinds of field theories that we will encounter.

We saw above that loop models can fruitfully be related to field theories with a continuous symmetry group, for example $O(n)$ in the simplest cases, and that a crucial trick is analytical continuation in parameters such as n. Such limits are reminiscent of the replica trick in disordered systems. Let us briefly recall this to see the similarities and differences.

As an illustration, take a real scalar field with an arbitrary potential V and quenched randomness $\Delta(x)$ in the mass term:

$$Z[\Delta] = \int \mathcal{D}\varphi\, e^{-S[\varphi,\Delta]}, \quad S[\varphi,\Delta] = \int \mathrm{d}x \left[(\nabla\varphi)^2 + V(\varphi) + \Delta\,\varphi^2 \right]. \tag{1.12}$$

$\Delta(x)$ is a random field, which is *not* integrated over in the partition function, and which we assume to have short-range correlations. For simplicity, we take the distribution of Δ to be Gaussian white noise.

The replica trick gives convenient expressions for disorder-averaged correlation functions, such as $\overline{\langle \varphi(x)\varphi(y)\rangle}$ (the overbar denotes the disorder average). Writing

$$\overline{\langle \varphi(x)\varphi(y)\rangle} = \lim_{n\to 0}\, \overline{Z[\Delta]^{n-1} \int \mathcal{D}\varphi\, e^{-S[\varphi,\Delta]} \varphi(x)\varphi(y)}, \tag{1.13}$$

we then forget that n is not a positive integer. $Z[\Delta]^{n-1}$ yields functional integrals over $(n-1)$ more 'replicas' of φ, so that altogether we have n fields which we collect into the vector $\boldsymbol{\varphi} = (\varphi^1, \ldots, \varphi^n)$:

$$\overline{\langle \varphi(x)\varphi(y)\rangle} = \lim_{n\to 0} \overline{\int \mathcal{D}\boldsymbol{\varphi}\, e^{-\sum_a S[\varphi^a,\Delta]} \varphi^1(x)\varphi^1(y)}. \tag{1.14}$$

Finally, averaging over Δ simply replaces $\sum_a S[\varphi^a, \Delta]$ with an effective action for the replicated fields,

$$\widetilde{S}[\boldsymbol{\varphi}] = \int \mathrm{d}x \left[(\nabla\boldsymbol{\varphi})^2 - \lambda\, (\boldsymbol{\varphi}^2)^2 + \sum\nolimits_a V(\varphi^a) \right], \tag{1.15}$$

where the variance of Δ determines λ. Taking into account the fact that the partition function associated with $\widetilde{S}$ is equal to one in the limit $n \to 0$, we may write the averages of correlators, and of products of correlators, in terms of the replicated fields—for example

$$\overline{\langle\varphi(x)\varphi(y)\rangle} = \lim_{n\to 0}\langle\varphi^1(x)\varphi^1(y)\rangle, \quad \overline{\langle\varphi(x)\varphi(y)\rangle^2} = \lim_{n\to 0}\langle\varphi^1(x)\varphi^2(x)\varphi^1(y)\varphi^2(y)\rangle.$$

The second formula suggests a heuristic interpretation of distinct replicas as separately equilibrated copies of the system, subject to the same disorder realisation.

Let us now compare with the 'replica-like' field theories for the loop models. To begin with, note that the symmetry group of Eq. 1.15 is in general the discrete group S_n (permutations) rather than a continuous one. The symmetry is promoted to $O(n)$ if we take the initial field theory to be free, $V(\varphi) \propto \varphi^2$. However, the potential for φ then has the 'wrong' sign, so we are still a long way from (say) de Gennes' description of the polymer, which corresponds to the case with positive λ.

These observations might seem to suggest there is little connection between the replica-like field theories for the loop models and those for disordered systems, but this is not always the case.

The quantum mechanics of free fermions in disordered potentials may often be described by replica sigma models for bosonic fields (for a review, see [70]) which have many similarities with the field theories that describe loop models. In the absence of interactions, Green's functions can be represented with path integrals for free fermionic fields (defined on space, rather than on space time). The sigma models appear after disorder-averaging the replicated free fermion action, making a Hubbard Stratonovich transformation with a (bosonic) matrix-valued field, integrating out the original replicated fermions, and finally making a saddle point approximation designed to isolate the important degrees of freedom. Their utility is in describing and classifying the universal statistical properties of the wavefunctions, in particular near Anderson transitions at which these wavefunctions become statistically scale invariant.

These sigma models are classified by the symmetries of the localisation problem and show a variety of critical points and critical phases. Many of these represent long-unsolved problems in 2D critical phenomena [70], with the most notorious example being the Pruisken sigma model for the quantum Hall transition [71]. In fact this field theory and the CP^{n-1} model both belong to a larger class of Grassmanian sigma models [72]. The Grassmanian is the manifold[12]

$$\frac{U(n)}{U(p)\times U(n-p)}. \tag{1.16}$$

Setting $p = 1$ gives CP^{n-1}, while the Pruisken sigma model is given by setting $n = 2p$ and then taking the limit $p \to 0$. In many ways it is analogous to the $n \to 1$ limit of the CP^{n-1} model. Both field theories permit the addition of a θ term, and

[12] This space can be parameterised by a Hermitian matrix constrained to have p eigenvalues equal to λ_1, and $n-p$ eigenvalues equal to λ_2, with λ_1 and λ_2 fixed and distinct. An equivalent parametrisation uses p distinct orthonormal n-component vectors, with a $U(p)$ gauge symmetry that rotates them within the p-dimensional subspace that they span. At $p = 1$, these two formulations correspond to Q in Eq. 1.10 and $\mathbf{z}$ in Eq. 1.9 respectively.

show critical behaviour at $\theta = \pi$ [11], expected to be described by 'logarithmic' conformal field theories with central charge zero [37–39].

The similarity between these field theories is not surprising if we recall from Sect. 1.3.2 that CP^0 describes the statistics of percolation cluster boundaries (on the lattice) and of level lines of random potentials (in the continuum), and that the latter yield a semiclassical picture for eigenstates at the quantum hall transition [73]. In general tunneling between level lines spoils this correspondence [74], so the Pruisken and CP^0 sigma models are in distinct universality classes. Remarkably, though, there is an exact correspondence between percolation/the CP^0 model (more precisely, the $CP^{1|1}$ model) and another localisation problem [1, 2, 75]. This is the spin quantum Hall transition [76–79], which is a quantum-Hall like transition for Bogoliubov quasiparticles in a disordered superconductor (in the symmetry classification of localisation problems, it is in class C rather than A). This correspondence also extends to 3D [2, 80, 81].

The Anderson critical points mentioned so far (quantum Hall and spin quantum Hall) appear at transitions between two insulating states. In some symmetry classes for disordered fermions 2D metallic phases are also allowed, and then metal-insulator transitions can be studied. In Chap. 5 we will explore classical analogues of such transitions.

As in the loop models, SUSY provides an alternative to the replica limit in noninteracting disordered systems [23, 82–84].[13] For generic interacting disordered systems, the SUSY alternative is not available. It is also important to emphasise that interacting disordered systems can show phenomena which are not (as far as we know) manifested in simple geometrical models or noninteracting disordered systems. For example, the permutational replica symmetry of the Ising spin glass—the Ising model with an appropriate distribution of ferromagnetic and antiferromangetic bonds—can be broken in extremely complex ways. The continuous[14] replica symmetries above can also be broken (in the loop models, this represents a proliferation of loops) but the symmetry-broken phase can usually be more simply understood by analogy with the integer n case.

[13] An additional complication in the SUSY sigma models for disordered fermions, in contrast to the loop models described above, is that generically (but not always) the target spaces are noncompact [70]. A SUSY sigma model with a noncompact target space that is related to both localisation and random walks is discussed in Ref. [85] (see also Ref. [86]).

[14] Geometrical models can also show discrete replica-like symmetries—in particular percolation is related to the $q \to 1$ state Potts model, as mentioned above. The percolating phase corresponds to the Potts ordered phase. Two-dimensional percolation is special, in that it can be mapped to a loop model by considering cluster boundaries, and thus also has a field theory description with a continuous replica-like symmetry.

1.9 This Thesis

Above, we began by reviewing a family of loop models that has been very important in the development of the geometrical approach to critical behaviour in 2D systems. One of the themes of the following chapters is the extension of the geometrical approach to situations in which it is less well developed: 3D or $(2 + 1)$D systems in Chaps. 2, 3, and 4, and 2D systems that lie outside the established paradigms in Chaps. 5 and 6. Another theme involves critical behaviour in CP^{n-1} and RP^{n-1} sigma models, especially (but not exclusively) in the replica limit, and the role played by various types of topological defects and textures in these field theories.

The following chapters are closely tied together by these themes (and we hope a unified picture of a family of universality classes emerges at the end) but they are motivated by a variety of different physical questions.

1.9.1 3D Loop Models and the CP^{n-1} *Model*

Chapter 2 begins with a class of models for directed loops in 3D. Certain of these models permit mappings to network models for Anderson localisation in symmetry class C [80, 81] and to 2D quantum magnets with $SU(n)$ symmetry. More importantly, they are prototypes for loop ensembles that occur in diverse contexts in 3D or $(2 + 1)$D statistical mechanics, and the lessons we learn from them can be readily generalised to various other problems formulated in terms of loops.

We give a simple mapping of the models to lattice CP^{n-1} models, and use these field theories to characterise the universal behaviour of geometrical observables both at continuous phase transitions and in phases with infinite loops—where, for example, the joint distribution of loop lengths is nontrivial. Motivated by a numerical analysis of the models, we study critical behaviour in the CP^{n-1} model in an expansion around $n = 2$ near four dimensions. We briefly show how to generalise our conclusions to some other three dimensional loop models, including models of undirected loops and models related to frustrated magnets of various kinds, by finding appropriate lattice mappings.

In Chap. 3 we show the connections that exist between the loop models and quantum magnets. We also clarify the role of hedgehog defects and imaginary contributions to the continuum action for the lattice CP^{n-1} models of Chap. 2, making contact with the literature on deconfined quantum critical points (at which the spin fractionalises into deconfined bosonic spinons, and where the appropriate description, the NCCP^{n-1} model, involves an emergent noncompact gauge field). We explain how to realise such critical points in the loop models. This chapter also reviews some background on the CP^{n-1} and NCCP^{n-1} models.

1.9.2 Fractal Geometry of Line Defects

Vortex lines are ubiquitous in random or disordered 3D systems. They show universal statistical properties on long scales, and can undergo geometrical phase transitions analogous to percolation transitions. Their fractal geometry has been studied numerically in diverse contexts, ranging from lattice magnets, through 'optical vortices' in laser speckle, to cosmic strings.

Despite this, and despite the importance of vortex defects in duality transformations in 3D, field theories that capture their statistical properties have not been available. In Chap. 4 we use mappings to lattice gauge theory to show that the CP^0 and RP^0 models provide the required descriptions. These field theories explain the numerical findings and allow us to classify 'geometrical' phase transitions in 3D disordered systems, showing that there are more possibilities than just the long-familiar one of percolation.

1.9.3 2D Loop Models with Crossings

2D loop models without crossings have various seductive features—exact results may be obtained from the Coulomb gas and other techniques, and some of these results are now being made rigorous with the machinery of SLE. This is perhaps why some dramatically different types of universal behaviour, which are possible when loops are allowed to cross, have been understood belatedly or not at all.

In the replica limit, it is possible for 2D sigma models to have 'Goldstone' phases in which the sigma model stiffness flows to infinity in the infrared. The beta function for the $O(n)$ sigma model shows that this happens when $n < 2$. Jacobsen, Read and Saleur argued that this Goldstone phase is what generically replaces the dense phase when crossings are allowed [9].

One of the aims of Chap. 5 is to pin down the universal behaviour in the Goldstone phase analytically for the first time, using RG for the sigma model to calculate various observables, and to compare these results with extensive numerical simulations.

The larger aim of this chapter is to reveal new types of critical behaviour in 2D loop models with crossings, which occur at transitions out of the Goldstone phase. We show that the field theories describing models with crossings are generically not $O(n)$ models: an additional $\mathbb{Z}_2$ gauge symmetry means that the appropriate descriptions are RP^{n-1} models. This leads to the possibility of phase transitions associated with the unbinding of $\mathbb{Z}_2$ vortex defects in the sigma model field. Unlike the Goldstone phase, where asymptotically exact results can be obtained by perturbative RG, no exact techniques are currently available for these critical points. However we give detailed numerical and approximate RG treatments.

These phase transitions may be regarded as classical analogues of metal-insulator or metal-topological insulator transitions. The analogy is particularly close for transitions in the symplectic symmetry class [87–91]. An open question is how far this analogy can be pushed.

1.9.4 Theta-Point Polymers

The collapse transition of a linear polymer in two dimensions is a fascinating subject in which many important issues remain mysterious. Chapter 6 tackles some of them, focussing on 2D models in which the polymer is allowed to cross itself. In particular, we consider the collapse transition for the well-studied 'interacting self-avoiding trail' (ISAT) model [92, 93]. There have been numerous conflicting claims about critical behaviour in this model, for which the correct field theory description was previously missing. Using the results of Chap. 5, we give a field theory treatment that resolves the confusion about exponents and explains the model's phase diagram. This also leads to the surprising conclusion that the ISAT collapse transition is an infinite-order multicritical point; we connect the fine-tuning that this implies with an enhanced symmetry which is inseparable from some of the model's useful (e.g. for numerics) features. In the light of this, we discuss RG flows for more general models for the Θ point, suggesting that exact critical exponents for fully generic models are still unknown.

References

1. I.A. Gruzberg, A.W.W. Ludwig, N. Read, Phys. Rev. Lett. **82**, 4524 (1999)
2. E.J. Beamond, J. Cardy, J.T. Chalker, Phys. Rev. B **65**, 214301 (2002)
3. H. Saleur, B. Duplantier, Phys. Rev. Lett. **58**, 2325 (1987)
4. B. Duplantier, H. Saleur, Phys. Rev. Lett. **59**, 539 (1987)
5. U. Wolff, Phys. Rev. Lett. **62**, 361 (1989)
6. H.G. Evertz, Adv. Phys. **52**, 1 (2003)
7. B. Nienhuis, Chap. 1, in *Phase Transitions and Critical Phenomena*, vol. 11, ed. by C. Domb, J. Liebowitz (Academic Press, Singapore, 1987)
8. O. Schramm, Israel J. Math. **118**, 221 (2001)
9. J.L. Jacobsen, N. Read, H. Saleur, Phys. Rev. Lett. **90**, 090601 (2003)
10. C.M. Fortuin, P.W. Kasteleyn, Physica **57**, 536 (1972)
11. N. Read, H. Saleur, Nucl. Phys. B **613**, 409 (2001)
12. A. Nahum, P. Serna, A.M. Somoza, M. Ortuño, Phys. Rev. B **87**, 184204 (2013)
13. J. Cardy, Ann. Physics **318**, 81 (2005)
14. B. Duplantier, Phys. Rep. **184**, 229 (1989)
15. J.L. Jacobsen, Chap. 14, in *Polygons, Polyominoes, and Polyhedra*, ed. by A.J. Guttmann (Springer, New York, 2009)
16. I. Gruzberg, J. Phys. A: Math. Gen. **39**, 12601 (2006)
17. E. Domany, D. Mukamel, B. Nienhuis, A. Schwimmer, Nucl. Phys. B **190**, 279 (1981)
18. B. Nienhuis, Phys. Rev. Lett. **49**, 1062 (1982)
19. H.W.J. Blöte, B. Nienhuis, Phys. Rev. Lett. **9**, 1372 (1994)
20. J. Kondev, C.L. Henley, Phys. Rev. Lett. **73**, 2786 (1994)
21. P.G. de Gennes, Phys. Lett. A **38**, 339 (1972)
22. K.G. Wilson, M.E. Fisher, Phys. Rev. Lett. **28**, 240 (1972)
23. A.J. McKane, Phys. Lett. A **76**, 22 (1980)
24. G. Parisi, N. Sourlas, J. Phys. Lett. **41**, L403 (1980)
25. C. Candu, J.L. Jacobsen, N. Read, H. Saleur, J. Phys. A **43**, 142001 (2010)
26. N. Read, H. Saleur, Nucl. Phys. B **777**, 263 (2007)

27. I. Affleck, Phys. Rev. Lett. **66**, 2429 (1991)
28. A. Nahum, J.T. Chalker, Phys. Rev. E **85**, 031141 (2012)
29. R. Shankar, G. Murthy, Phys. Rev. B **72**, 224414 (2005)
30. J. Kondev, Phys. Rev. Lett. **78**, 4320 (1997)
31. J.L. Jacobsen, J. Kondev, J. Stat. Phys **96**, 21 (1999)
32. J.L. Jacobsen, J. Kondev, Nucl. Phys. B **532**, 635 (1998)
33. P.J. Flory, Nobel lecture (1974)
34. D.J. Thouless, J. Phys. C **8**, 1803 (1975)
35. G. Parisi, N. Sourlas, Phys. Rev. Lett. **46**, 871 (1981)
36. V. Gurarie, Nucl. Phys. B **410**, 535 (1993)
37. A. Gainutdinov, J.L. Jacobsen, N. Read, H. Saleur, R. Vasseur, J. Phys. A: Math. Theor. **46**, 494012 (2013)
38. J. Cardy, J. Phys. A: Math. Theor. **46**, 494001 (2013)
39. V. Gurarie, J. Phys. A: Math. Theor. **46**, 494003 (2013)
40. J.L. Jacobsen, N. Read, H. Saleur, Phys. Rev. Lett. **93**, 038701 (2004)
41. A. Coniglio, N. Jan, I. Majid, H.E. Stanley, Phys. Rev. B **35**, 3617 (1987)
42. M. Kardar, Y.C. Zhang, Phys. Rev. Lett. **58**, 20872090 (1987)
43. D.C. Brydges, J.Z. Imbrie, Ann. Math **158**, 1019 (2003)
44. J. Cardy, (2003), arXiv:0302495
45. R.K. Kaul, R.G. Melko, A.W. Sandvik, Annu. Rev. Condens. Matter Phys. **4**, 179 (2013)
46. D.S. Rokhsar, S.A. Kivelson, Phys. Rev. Lett. **61**, 2376 (1988)
47. R. Moessner, S.L. Sondhi, Phys. Rev. Lett. **86**, 1881 (2001)
48. A. Kitaev, Ann. Phys. (N.Y.) **303**, 2 (2003)
49. M. Freedman et al., Ann. Phys. (N.Y.) **310**, 428 (2004)
50. M. Freedman, C. Nayak, K. Shtengel, Phys. Rev. Lett. **94**, 147205 (2005)
51. M. Freedman, C. Nayak, K. Shtengel, Phys. Rev. B **78**, 174411 (2008)
52. M. Troyer, S. Trebst, K. Shtengel, C. Nayak, Phys. Rev. Lett. **101**, 230401 (2008)
53. M.A. Levin, X.G. Wen, Phys. Rev. B **71**, 045110 (2005)
54. P. Fendley, Ann. Phys. (N.Y.) **323**, 3113 (2008)
55. P. Fendley, S.V. Isakov, M. Troyer, Phys. Rev. Lett. **110**, 260408 (2013)
56. K. Damle, D. Dhar, K. Ramola, Phys. Rev. Lett. **108**, 247216 (2012)
57. A. Fabricio Albuquerque, F. Alet, R. Moessner, Phys. Rev. Lett. **109**, 147204 (2012)
58. E. Witten, Nucl. Phys. B **149**, 285 (1979)
59. A. D'Adda, M. Lüscher, P. Di Vecchia, Nucl. Phys. B **146**, 63 (1978)
60. C. Candu, V. Mitev, T. Quella, H. Saleur, V. Schomerus, J. High Energy Phys. **02**, 015 (2010)
61. O.I. Motrunich, A. Vishwanath, Phys. Rev. B **70**, 075104 (2004)
62. T. Senthil, L. Balents, S. Sachdev, A. Vishwanath, M.P.A. Fisher, Phys. Rev. B **70**, 144407 (2004)
63. T. Senthil, A. Vishwanath, L. Balents, S. Sachdev, M.P.A. Fisher, Science **303**, 1490 (2004)
64. S. Powell, J.T. Chalker, Phys. Rev. Lett. **101**, 155702 (2008)
65. D. Charrier, F. Alet, P. Pujol, Phys. Rev. Lett. **101**, 167205 (2008)
66. S. Powell, J.T. Chalker, Phys. Rev. B **80**, 134413 (2009)
67. G. Chen, J. Gukelberger, S. Trebst, F. Alet, L. Balents, Phys. Rev. B **80**, 045112 (2009)
68. P.A. Lebwohl, G. Lasher, Phys. Rev. A **6**, 426 (1972)
69. J. Cardy, *Scaling and Renormalization in Statistical Physics* (Cambridge University Press, Cambridge, 1996)
70. F. Evers, A.D. Mirlin, Rev. Mod. Phys. **80**, 1355 (2008)
71. A.M.M. Pruisken, Nucl. Phys. B **235**, 277 (1984)
72. I. Affleck, Nucl. Phys. B **257**, 397 (1985)
73. S.A. Trugman, Phys. Rev. B **27**, 7539 (1983)
74. J.T. Chalker, P.D. Coddington, J. Phys. C. **21**, 2665 (1988)
75. A.D. Mirlin, F. Evers, A. Mildenberger, J. Phys. A: Math. Gen. **36**, 3255 (2003)
76. T. Senthil, M.P.A. Fisher, L. Balents, C. Nayak, Phys. Rev. Lett. **81**, 4704 (1998)
77. R. Bundschuh, C. Cassanello, D. Serban, M.R. Zirnbauer, Phys. Rev. B **59**, 4382 (1999)

78. T. Senthil, J.B. Marston, M.P.A. Fisher, Phys. Rev. B **60**, 4245 (1999)
79. V. Kagalovsky, B. Horovitz, Y. Avishai, J.T. Chalker, Phys. Rev. Lett. **82**, 3516 (1999)
80. M. Ortuño, A.M. Somoza, J.T. Chalker, Phys. Rev. Lett. **102**, 070603 (2009)
81. J. Cardy, in *50 Years of Anderson Localization*, ed. by E. Abrahams (World Scientific, Singapore, 2010)
82. K.B. Efetov, Adv. Phys. **32**, 53 (1983)
83. K. Efetov, *Supersymmetry in Disorder and Chaos* (Cambridge University Press, Cambridge, 1997)
84. F. Haake, *Quantum Signatures of Chaos* (Springer, Berlin, 1991)
85. M. Disertori, T. Spencer, M.R. Zirnbauer, Commun. Math. Phys. **300**, 435 (2010)
86. Y. Ikhlef, J. Jacobsen, H. Saleur, Nucl. Phys. B **789**, 483 (2008)
87. L. Fu, C.L. Kane, Phys. Rev. Lett. **109**, 246605 (2012)
88. Y. Asada, K. Slevin, T. Ohtsuki, Phys. Rev. Lett. **89**, 256601 (2002)
89. P. Markos, L. Schweitzer, J. Phys. A **39**, 3221 (2006)
90. A. Mildenberger, F. Evers, Phys. Rev. B **75**, 041303 (2007)
91. H. Obuse, A.R. Subramaniam, A. Furusaki, I.A. Gruzberg, A.W.W. Ludwig, Phys. Rev. B **82**, 035309 (2010)
92. J.W. Lyklema, J. Phys. A. Math. Gen. **18**, L617 (1985)
93. A.L. Owczarek, T. Prellberg, J. Stat. Phys. **79**, 951 (1995)

Chapter 2
Completely Packed Loop Models

2.1 Introduction

We begin with a class of three-dimensional loop models that show transitions between two types of phase—one with infinite loops, and one with only short loops. In particular, we focus on a family of completely packed models that may be viewed as discretisations of sigma models with target space CP^{n-1} (and RP^{n-1}). These loop models capture the universal behaviour of a surprising variety of classical and quantum systems—including $(2+1)$-dimensional $SU(n)$ antiferromagnets (generalising the connection for $(1+1)$D spin chains reviewed in Sect. 1.4.2), Anderson localisation in symmetry class C [1–4], line defects in random media [5], and certain models for polymers.

We will return to some of these connections in later chapters. Our aims here are to establish the continuum descriptions of the loop models, and to use these descriptions to characterise their phase transitions and the universal properties of geometrical observables. A short summary was given in Sect. 1.9.1, so we move immediately to concrete models.

2.2 Definitions

Take any four-coordinated, directed lattice on which two links enter and two links exit each node. On such a lattice we may generate a configuration of oriented loops by picking a pairing of incoming and outgoing links at each node. We assign weight p to one such pairing and weight $1-p$ to the other, so that if N_p and N_{1-p} are the numbers of nodes where the weight p and $1-p$ pairings are followed, the total weight associated with the nodes in a configuration $\mathcal{C}$ is

$$W_{\mathcal{C}} = p^{N_p}(1-p)^{N_{1-p}}. \tag{2.1}$$

A. Nahum, *Critical Phenomena in Loop Models*, Springer Theses,
DOI 10.1007/978-3-319-06407-9_2

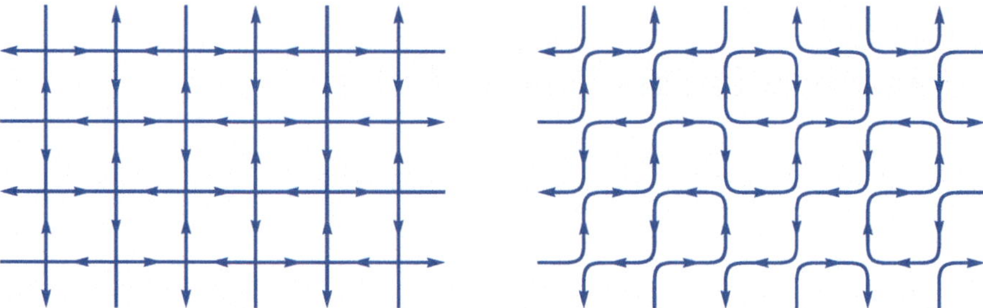

Fig. 2.1 The 2D 'L' lattice and a configuration obtained by resolving the nodes

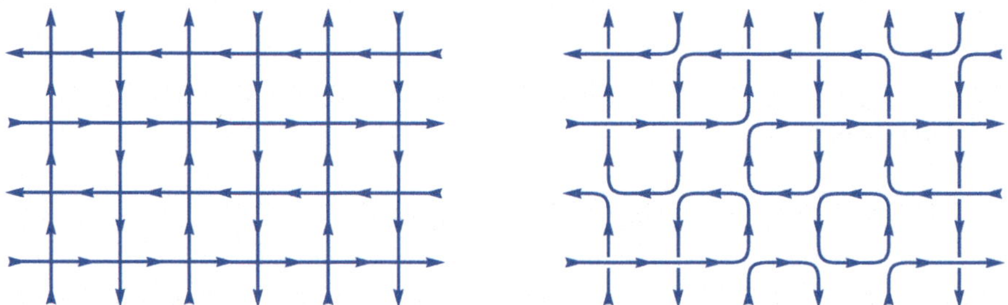

Fig. 2.2 The 2D Manhattan lattice and a configuration obtained by resolving the nodes

In addition we assign the loops a fugacity n. Denoting the number of loops by $|\mathcal{C}|$, the partition function is

$$Z_{\text{loops}} = \sum_{\mathcal{C}} n^{|\mathcal{C}|} W_{\mathcal{C}}. \tag{2.2}$$

The fugacity may also be generated by summing over n possible colours for each loop.

In the special case $n = 1$, the nodes are independent and the parameter p is simply the probability of a given node configuration. The problem is then akin to percolation: while the local degrees of freedom have trivial correlation functions, geometrical observables show nontrivial behaviour.

Various models can be constructed on the above lines, depending on the choice of lattice and the assignment of node parameters. The best-studied case is the loop model on the two-dimensional L lattice (Fig. 2.1), where weight p is associated with right-turning nodes and weight $1 - p$ with left-turning nodes (Fig. 2.3, left). This model may be mapped to the n^2-state Potts model via the Fortuin-Kasteleyn representation of the latter, or to bond percolation when $n = 1$. Another natural model in two dimensions is that on the Manhattan lattice (Fig. 2.2) where all the links on a given line point in the same direction, and where weight p is assigned to nodes at which the loops cross and weight $1 - p$ to those at which they turn (Fig. 2.3, right). See Chap. 5 for further discussion of models on these lattices.

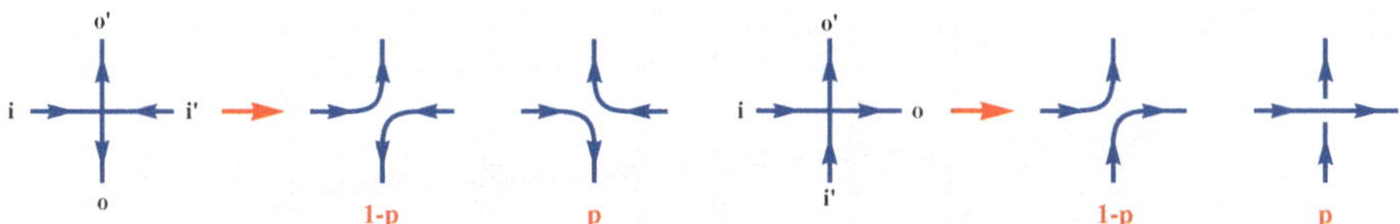

Fig. 2.3 *Left* possible decompositions of a node of the 2D or 3D L lattice (the labelling of links used in Eq. 2.6 is also shown). *Right* decompositions of a node of the 2D Manhattan or 3D K lattice

In this chapter we consider two lattices that allow us to explore the range of critical phenomena possible in three dimensions. These are Cardy's 3D L lattice [4], and a variant, the K lattice (which is similar to the 3D Manhattan lattice of Ref. [4]) that shares the same underlying undirected lattice but has different link orientations. Previous numerical work at $n = 1$ considered a model on the diamond lattice [3]. Much of the discussion will however be independent of the choice of lattice, and even the restriction to four-coordinated nodes can be dropped.

2.2.1 The 3D L and K Lattices

Following Ref. [4], consider two interpenetrating cubic lattices of lattice spacing two, $C_1 = (2\mathbb{Z})^3$ and $C_2 = (2\mathbb{Z} + 1)^3$, displaced from each other by the vector $(1, 1, 1)$. The desired lattice of coordination number four is formed by the intersections of the *faces* of C_1 and C_2—see Fig. 2.4. The 3D L lattice is obtained by orientating the links of this lattice so that each node resembles Fig. 2.3 (left) up to a rotation.

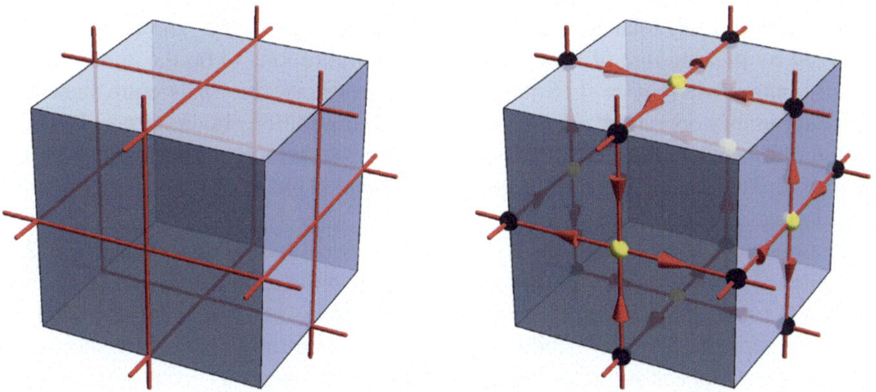

Fig. 2.4 *Left* a cube of C_1, with the lines of intersection with C_2 marked in *red*. These lines form the links of the L and K lattices. *Right* with the orientations corresponding to the L lattice added. The two sublattices of nodes, A and B, are marked in *yellow* and *black* respectively

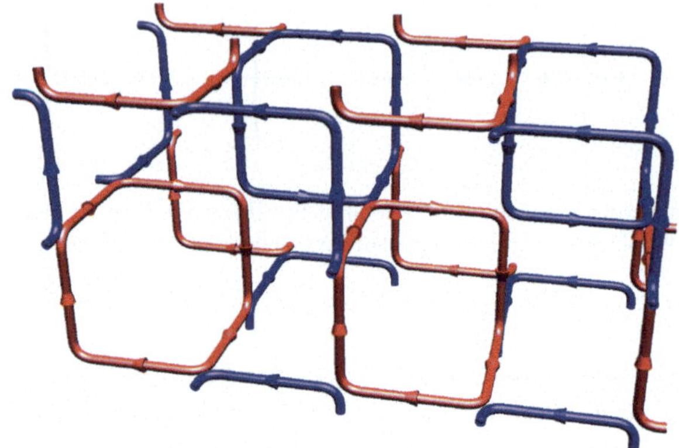

Fig. 2.5 *Left* configuration on the K lattice at $p = 0$

This choice is unique up to reversal of all orientations, and the L lattice has cubic symmetry (some symmetry operations include a reversal of the orientations). For the K lattice, we orient links so that every node resembles Fig. 2.3 (right), and so that nearest-neighbour parallel axes, which are at a distance of $\sqrt{2}$, are oppositely oriented. Again this choice is essentially unique (the eight choices of link orientations are related by reflections), but the resulting oriented lattice is anisotropic.

The K lattice is designed to ensure the existence of both extended and localised phases (phases with and without infinite loops, respectively) for all values of n. The node weights are fixed by Fig. 2.3: the loops reduce to infinite straight lines when $p = 1$, while at $p = 0$ all loops have the minimal length possible on these lattices, which is six. This $p = 0$ configuration is shown in Fig. 2.5. The phase diagram obtained numerically for the K lattice is shown in Fig. 2.6.

While the K lattice only permits a single packing of loops of length six, the L lattice permits four, related to each other by lattice symmetry. The node weights are fixed by picking one of these short-loop configurations to be the configuration which obtains when $p = 1$. Since the exchange $p \to 1 - p$ is equivalent to a lattice symmetry operation, another short-loop configuration obtains at $p = 0$. These configurations provide adequate caricatures of the short loop phases that exist for p close to zero or one (see Fig. 2.6); however an extended phase appears at intermediate p, so long as n is not too large.

The nodes of the L lattice form two sublattices, A and B, as shown in Fig. 2.4. (The sites of A lie at the centres of faces of C_1 and at the centres of edges of C_2, and vice versa for B. The total number of nodes is $3L^3/4$ in an $L \times L \times L$ system.) If we introduce separate node probabilities p_A and p_B, then the four distinct short-loop configurations appear at the four corners of the phase diagram (given by $p_A = \pm 1$, $p_B = \pm 1$). The weight (2.1) corresponds to the line $p_A = p_B = p$.

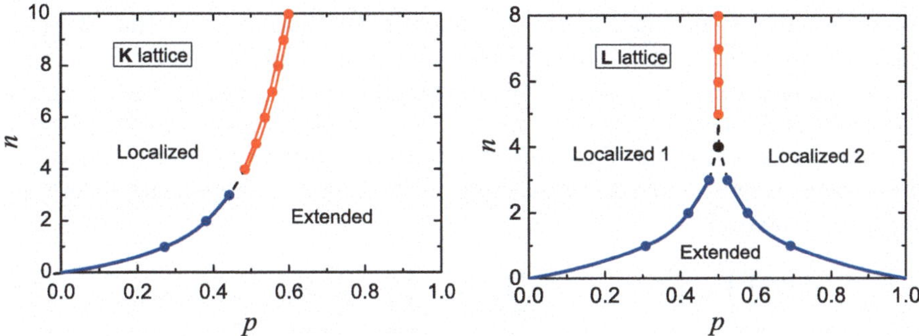

Fig. 2.6 Phase diagrams for the K and L lattices. *Red lines* indicate first order transitions and *blue lines* indicate those that appear to be continuous (*Note* on the L lattice, $p = 1/2$, $n = 4$ lies within the extended phase.)

2.3 Lattice Field Theories

The loop models may be related to lattice 'magnets' for spins located on the links l of the lattice. Neglecting for now complications associated with replicas or supersymmetry (and taking n to be a positive integer), these spins are n-component complex vectors $\mathbf{z}_l$:

$$\mathbf{z}_l = (z_l^1, \ldots, z_l^n), \quad \mathbf{z}_l^\dagger \mathbf{z}_l = n. \tag{2.3}$$

The action for these degrees of freedom will be chosen so that a graphical expansion generates the sum over loop configurations defining Z_{loops}. This leads to a $U(1)$ gauge symmetry,

$$\mathbf{z}_l \sim e^{i\phi_l}\mathbf{z}_l, \tag{2.4}$$

implying that the spins live on CP^{n-1} (defined in Sect. 1.7).

The required action may be written as a sum of contributions from nodes. Letting the trace 'Tr' stand for an integral over the fixed-length vectors $\mathbf{z}$, normalised so $\mathrm{Tr}\ 1 = 1$, we write

$$Z = \mathrm{Tr}\ \exp\left(-\sum_{\text{nodes}} S_{\text{node}}\right). \tag{2.5}$$

To define S_{node}, label the links at a node as in Fig. 2.3, with the weight p pairing being $i \to o$, $i' \to o'$ and the weight $1-p$ pairing being $i \to o'$, $i' \to o$. Then

$$\exp\left(-S_{\text{node}}\right) = p\,(\mathbf{z}_o^\dagger \mathbf{z}_i)(\mathbf{z}_{o'}^\dagger \mathbf{z}_{i'}) + (1-p)\,(\mathbf{z}_o^\dagger \mathbf{z}_{i'})(\mathbf{z}_{o'}^\dagger \mathbf{z}_i). \tag{2.6}$$

The total action is invariant under the gauge transformation (2.4)—the phase ϕ_l cancels between the terms for the two nodes adjacent to link l.

The two terms in $e^{-S_{\text{node}}}$ are in correspondence with the two possible configurations of the node in the loop model. This leads to a simple expression for the partition function in terms of loop configurations $\mathcal{C}$, obtained by writing the Boltzmann weight as a product over nodes and then expanding out this product. Denoting the links lying on a given loop by $1, \ldots, \ell$,

$$\text{Tr} \prod_{\text{nodes}} \exp\left(-S_{\text{node}}\right) = \sum_{\mathcal{C}} W_{\mathcal{C}} \prod_{\text{loops in } \mathcal{C}} \text{Tr}\ (\mathbf{z}_1^\dagger \mathbf{z}_2) \ldots (\mathbf{z}_\ell^\dagger \mathbf{z}_1). \tag{2.7}$$

The integrals over the $\mathbf{z}$s are performed using Tr $z_l^\alpha \bar{z}_l^\beta = \delta^{\alpha\beta}$, leading to a single free colour index for each loop as in Eq. 1.6:

$$Z = \sum_{\mathcal{C}} \sum_{\substack{\text{loop} \\ \text{colours}}} W_{\mathcal{C}} = Z_{\text{loops}}. \tag{2.8}$$

This establishes the desired correspondence between the loop models and models with local 'magnetic' degrees of freedom. We now make the simplest conjectures for the models' continuum descriptions, taking account of the $SU(n)$ global and $U(1)$ gauge symmetries of (2.5).

Note that the lattice action in Eq. 2.5 is not necessarily real. This is not an obstacle to taking the continuum limit (recall for example that complex lattice actions are the norm in quantum problems). In fact it can be an important feature of loop models of this type, leading to imaginary terms in the continuum Lagrangian that are associated with spin configurations of nontrivial topology. These subtleties will not play a role in this chapter, but it will be necessary to consider them more carefully in the next.

2.4 Continuum Descriptions

Let us exchange the vector $\mathbf{z}$, which is a redundant parametrisation of CP^{n-1}, for the gauge-invariant matrix $Q^{\alpha\beta} = z^\alpha \bar{z}^\beta - \delta^{\alpha\beta}$. In addition to being Hermitian and traceless, Q obeys the nonlinear constraint

$$(Q+1)^2 = n\,(Q+1). \tag{2.9}$$

In the continuum we may either retain this constraint, giving the CP^{n-1} sigma model,

$$\mathcal{L}_\sigma = \frac{1}{2g} \text{tr}\ (\nabla Q)^2 \qquad\qquad (\&\text{ constraint on } Q) \tag{2.10}$$

or we may use a soft-spin formulation in which Q is an arbitrary traceless Hermitian matrix:

$$\mathcal{L}_{\text{soft}} = \text{tr } (\nabla Q)^2 + t \text{ tr } Q^2 + g \text{ tr } Q^3 + \lambda \text{ tr } Q^4 + \lambda'(\text{tr } Q^2)^2. \tag{2.11}$$

We will see shortly that the extended phase of the loop models corresponds to the ordered phase of the CP^{n-1} model, with $\langle Q \rangle \neq 0$, and short-loop phases correspond to disordered phases.

In using the language of the sigma model, we should bear in mind that in three dimensions the field $Q(x)$ can have pointlike topological defects (hedgehogs) associated with the second homotopy group of CP^{n-1}:

$$\pi_2(CP^{n-1}) = \mathbb{Z}. \tag{2.12}$$

These play an important role in the vicinity of the critical point [6–8], and they proliferate in disordered phase; they are of course irrelevant in the ordered phase. In order to accomodate them, the sigma model constraint must be relaxed in the defect core, and the regularisation-dependent physics in the core then determines a finite fugacity for defects. The condensed formulation of Eq. 2.10, with only the single parameter g, is therefore slightly misleading.

The manifold CP^1 is simply the sphere, so at the special value $n = 2$ the above field theories reduce to the sigma model and soft spin incarnations of the $O(3)$ model. At this value of n, the cubic term in $\mathcal{L}_{\text{soft}}$ vanishes, and there is only one quartic term (the latter also holds at $n = 3$). The $O(3)$ spin S is related to Q via the Pauli matrices, $Q = \frac{1}{\sqrt{2}}\sigma^i S^i$, giving

$$\mathcal{L}_{\text{soft}} = (\nabla S)^2 + t\, S^2 + u\,(S^2)^2 \tag{2.13}$$

with $u = \lambda' + \lambda/2$.

This description leads us to expect continuous transitions in the $O(3)$ universality class for the $n = 2$ models. Continuous transitions are also expected for $n < 2$. For $n > 2$, the naive expectation is a first order transition, as a result of the cubic term in $\mathcal{L}_{\text{soft}}$. However we will argue in Sect. 2.8.2 that fluctuations can invalidate this mean field prediction when the spatial dimension is less than four—numerical work is thus required to determine what happens in 3D.

2.5 Correlation Functions

The natural correlation functions in the loop models are the probabilities of various geometrically-defined events. The basic ones are the l-leg 'watermelon' correlators $G_l(x, y)$, defined as the probability that l distinct strands of loop connect x with y (here we use a continuum notation). If all loops close, the number l must of course be even. On the lattice, G_2 can be taken to be the probability that two links lie on the same loop, and G_4 the probability that two nodes are connected by four strands.

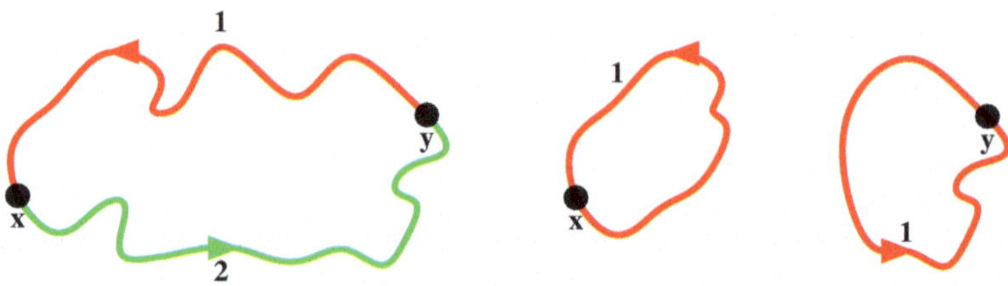

Fig. 2.7 *Left* $Q^{12}(x)$ is a two-leg operator, absorbing an incoming strand of colour 1 and emitting an outgoing one of colour 2. This allows us to form $G_2(x, y)$. *Right* with only a single colour at our disposal (when $n = 1$) we cannot control the topology of the strands passing through x and y

These geometrical observables may be expressed as correlation functions of Q. At a formal level, the relationships follow straightforwardly from the graphical expansion of Sect. 2.3, but they may also be easily guessed by thinking of the loops as worldlines of $\mathbf{z}$ quanta, i.e. by viewing one of the spatial directions as the imaginary time direction for a quantum problem (as in Sect. 1.3.2). These worldlines come in n colours, one for each component of $\mathbf{z}$, and since z^α is a complex field they carry an orientation distinguishing particles from antiparticles.

In particular, the operator $Q_l^{\mu\nu} = z_l^\mu \bar{z}_l^\nu$ with $\mu \neq \nu$ absorbs an incoming strand (worldline) of colour μ and emits an outgoing one of colour ν (Fig. 2.7). More precisely, the effect of this operator on the graphical expansion is to force the loop passing through link l to change colour from μ to ν there. This follows from the fact that for link l, the integral appearing in Eq. 2.7 is modified by the presence of the operator from $\text{Tr}\ z_l^\alpha \bar{z}_l^{\alpha'} = \delta^{\alpha\alpha'}$ to $\text{Tr}\ z_l^\alpha \bar{z}_l^{\alpha'} Q_l^{\mu\nu} = A\,\delta^{\alpha\nu}\delta^{\alpha'\mu}$ (for $\mu \neq \nu$), with $A = n/(n+1)$.

Assume that all loops close, i.e. that there are no open strands escaping to infinity or ending on the boundary (we will relax this in Sect. 2.7). Then the correlator $\langle Q_l^{12} Q_{l'}^{21} \rangle$ receives contributions only from configurations in which l and l' are joined (by a loop with one arm of colour 1 and one arm of colour 2, as in Fig. 2.7), and is proportional to the 2-leg correlator:

$$G_2(l, l') = \frac{n}{A^2} \langle Q_l^{12} Q_{l'}^{21} \rangle. \tag{2.14}$$

The factor of n compensates for the fact that there is no sum over colour indices for the loop passing through l and l'. For convenience, we use here the off-diagonal elements of Q to write geometrical correlators, but all Q correlators can be expressed in terms of loops; for example $\langle \text{tr}\ Q_l Q_{l'} \rangle = (n-1)A\, G_2(l, l')$. We may think of the diagonal components of Q as operators which measure the colour of a link.

In the continuum, the higher watermelon correlators may be written[1]:

$$G_{2k}(x,y) \propto \left\langle [Q^{12}(x)]^k [Q^{21}(y)]^k \right\rangle. \tag{2.16}$$

The correlators considered so far involve only two distinct spin indices, so can be written down as long as $n \geq 2$. More complex correlation functions may require the use of more indices. For example, in Sect. 2.9 we will require the probability that a single loop passes through all the links $l_1, \ldots, l_m$, in that order; this can be written down only if $n \geq m$:

$$G(l_1, \ldots, l_m) = \frac{n}{A^m} \left\langle Q^{12}_{l_1} Q^{23}_{l_2} \ldots Q^{m1}_{l_m} \right\rangle. \tag{2.17}$$

2.6 Replicas and Supersymmetry

These formulas highlight a potential problem: the complexity of the geometrical correlation functions we can represent is limited by the number n of spin components at our disposal, which in turn is set by the loop fugacity. The problem is particularly acute at $n = 1$, when we cannot represent any correlation functions at all—indeed the manifold CP^{n-1} degenerates to a point when $n = 1$, and the matrix Q is identically zero.

As mentioned in Sect. 1.3.1, we can resolve this problem either using replica-like limits—for example, we can write the 2-leg correlator in the $n = 1$ loop model as $G_2(x,y) \propto \lim_{n\to 1}\langle Q^{12}(x) Q^{21}(y)\rangle$—or using SUSY. For most of our purposes these formulations are interchangeable, and it is easy to translate between them. We will mostly use the language of replicas, for presentational simplicity and because it allows us to consider noninteger loop fugacities, which will occasionally be useful. The SUSY formulation is useful for certain tasks; it is also reassuring to know that there is a more rigorous formulation of the field theories which avoids ad hoc analytical continuations.

In the SUSY approach, we augment the lattice field theory with additional Grassman degrees of freedom [9–12]. Every occurence of a $\mathbf{z}$ in (2.5) is replaced by a supervector $\boldsymbol{\psi}$ with $n+k$ commuting and k anticommuting components:

$$\boldsymbol{\psi} = (z^1, \ldots, z^{n+k}, \chi^1, \ldots, \chi^k). \tag{2.18}$$

[1] It is sometimes useful to gather operators into irreducible representations of $SU(n)$. The 4-leg operators form a representation of dimension $n^2(n-1)(n+3)/4$, given (in the sigma model) by the tensor (symmetric in α, α' and in β, β')

$$X^{\alpha\alpha'\beta\beta'} = z^\alpha z^{\alpha'} \bar{z}^\beta \bar{z}^{\beta'} - \frac{\delta^{\alpha\beta} z^{\alpha'} \bar{z}^{\beta'} + 3 \text{ other terms}}{n+2} + \frac{\delta^{\alpha\beta}\delta^{\alpha'\beta'} + \delta^{\alpha\beta'}\delta^{\alpha'\beta}}{(n+1)(n+2)} \tag{2.15}$$

The vector $\boldsymbol{\psi}^\dagger$ involves independent Grassman variables $\bar{\chi}$. Again, we enforce $\boldsymbol{\psi}^\dagger \boldsymbol{\psi} = n$; concretely, we may take this to mean that the integral over $\boldsymbol{\psi}$ defining the trace Tr is proportional to $\int \mathrm{d}\boldsymbol{\psi} \mathrm{d}\boldsymbol{\psi}^* \delta\left(\boldsymbol{\psi}^\dagger \boldsymbol{\psi} - n\right)$, where the delta function is defined by expanding in the Grassman variables. See Ref. [13, 14] for reviews of Grassman integration.

In the expansion of the Boltzmann weight in terms of loop configurations (Eq. 2.7), the integrals for a single loop in the graphical expansion become:

$$\mathrm{Tr}\ (\boldsymbol{\psi}_1^\dagger \boldsymbol{\psi}_2) \ldots (\boldsymbol{\psi}_\ell^\dagger \boldsymbol{\psi}_1) = \mathrm{Tr}\ \bar{\psi}_1^{\alpha_1} (\psi_2^{\alpha_1} \bar{\psi}_2^{\alpha_2}) \ldots (\psi_\ell^{\alpha_{l-1}} \bar{\psi}_\ell^{\alpha_\ell}) \psi_1^{\alpha_\ell}. \tag{2.19}$$

As before, the formula $\mathrm{Tr}\ \psi_l^\alpha \bar{\psi}_l^\beta = \delta^{\alpha\beta}$ forces all the spin indices to be equal, say to α. However a reordering of the ψs is required before using this formula, and this gives a minus sign if ψ^α is fermionic. This leads to a cancellation between bosonic and fermionic colours, so that the sum over α gives the desired loop fugacity, $(n+k) - k = n$, irrespective of k.

While the partition function and its loop expansion are independent of k, increasing k gives access to more operators. Geometrical correlators may be constructed as in the previous section (if we use only bosonic components the formulas are unchanged).

The field $Q_l = \boldsymbol{\psi}_l \boldsymbol{\psi}_l^\dagger - 1$ is now a supermatrix with both bosonic and fermionic blocks, and lives on complex projective superspace, denoted $\mathrm{CP}^{n+k-1|k}$. Correspondingly the SUSY version of $\mathcal{L}_\sigma$ is termed the $\mathrm{CP}^{n+k-1|k}$ sigma model. We will return to this field theory briefly in Chap. 3. Both the lattice Boltzmann weight and $\mathcal{L}_\sigma$ are invariant under 'rotations' of $\boldsymbol{\psi}$ which preserve $\boldsymbol{\psi}^\dagger \boldsymbol{\psi}$ (superunitary tranformations).

2.7 Phases of the CP^{n-1} Model and Loops

When the CP^{n-1} spins are disordered, correlators such as G_2 decay exponentially ($G_2(r) \sim r^{-1} e^{-r/\xi}$), indicating that long loops are exponentially suppressed. This corresponds to the short-loop phase. On approaching a critical point, the correlation length ξ—a measure of the typical linear size of a loop—diverges as $\xi \sim |p - p_c|^{-\nu}$.

Precisely at criticality, the anomalous dimension η of the field Q determines the decay of $G_2(x, y)$,

$$G_2(x, y) \sim |x - y|^{-(1+\eta)}, \tag{2.20}$$

as well as the fractal dimension of a loop (via a simple scaling relation [15, 16])

$$d_f = \frac{5 - \eta}{2}, \tag{2.21}$$

and the probability $P_{\rm link}(\ell)$ that the loop passing through a given link is of length ℓ:

$$P_{\rm link}(\ell) \sim \ell^{1-\tau}, \qquad \tau = \frac{11-\eta}{5-\eta}. \tag{2.22}$$

For the critical points discussed in this chapter, η is close to the mean field value $\eta = 0$, so the loops have a fractal dimension close to $5/2$. They are thus more compact than Brownian paths. (Unlike, say, a self-avoiding walk, or the loops which appear in the high temperature expansion of the critical Ising model, which both have fractal dimensions less than two.)

When the spins are ordered, $G_2(x, y)$ becomes long-ranged, indicating the appearance of 'infinite' loops. This is the extended phase. A basic consequence of the CP^{n-1} description is that individual loops are Brownian in this phase, having for example a fractal dimension equal to two. However not all universal properties can be understood by thinking about simple random walks, as we will see in Sect. 2.9.

In the limit of large system size L, there is a sharp distinction between 'finite' and 'infinite' loops, but the nature of the latter depends on whether curves are allowed to terminate at the boundary or whether only closed loops are allowed (as in a system with reflecting or periodic boundary conditions). In the former case the 'infinite' strands end on the boundary, and have length of order L^2 as a result of the fractal dimension being two. When all loops are closed, the 'infinite' loops instead have a length of order L^3: although the fractal dimension of a segment of radius $\lesssim L$ is still two, the number of times a given 'infinite' loop crosses the system is of order L. In this case their joint length distribution shows an interesting universal structure. We will describe this in Sect. 2.9—here we briefly discuss the definition of k-leg correlation functions in the extended phase.

Consider the case where strands can end on the boundary. Microscopically, we have dangling links on the boundary: in the lattice field theory (2.5) we set the spins on these links equal to a fixed vector $\mathbf{z} = \mathbf{e}$ ($\mathbf{e}$ is taken to be real). The graphical expansion then goes through as before, except that a strand of colour α ending on the boundary incurs weight e^{α}. This allows us to separate contributions to correlators associated with finite strands from those associated with infinite strands, i.e. those which end on the boundary.

The boundary condition $\mathbf{z} = \mathbf{e}$ fixes the direction of symmetry breaking in the CP^{n-1} model: we have $\langle Q(x)\rangle = Q_0\left(\mathbf{e}\mathbf{e}^\dagger - 1\right)$, where Q_0 determines the strength of long range order. Firstly, set $\mathbf{e} = (1, 1, \ldots, 1)$: then the expectation value $\left\langle Q^{12}(x)\right\rangle$ receives contributions only from configurations in which the two strands emanating from x terminate on the boundary (since the two legs have different colour, the loop cannot close in the bulk). Therefore Q_0 is proportional to the probability that a given link lies on an infinite loop. On approaching the critical point, this probability scales with the order parameter exponent β, $Q_0 \sim |p - p_c|^\beta$.

Now consider the separation dependence of the l-leg correlation functions. If we define $G_l^{\rm finite}(x, y)$ to be the probability that x and y are connected by l *finite* strands, l can be either odd or even (unlike at the critical point and in the short loop phase).

Free field theory for the Goldstone modes leads to simple Brownian exponents for these correlators:

$$G_l^{\text{finite}}(x, y) \sim |x - y|^{-l}. \tag{2.23}$$

Each factor of $1/\text{distance}$ can be viewed as the probability that a random walker who starts at one of the points happens to visit the other. It follows from Eq. 2.23 that the curves' fractal dimension is two and the length distribution of finite loops scales as in (2.22) with $\tau = 5/2$.

To see where (2.23) comes from, set $\mathbf{e} = (n, 0, \ldots, 0)$ and write Q in terms of $(n-1)$ complex Goldstone modes $\boldsymbol{\phi}$ (we neglect the massive Higgs mode)

$$Q(x) \propto \hat{\mathbf{z}}\hat{\mathbf{z}}^\dagger(x) - 1, \quad \hat{\mathbf{z}}(x) = \left(\sqrt{1 - |\boldsymbol{\phi}(x)|^2}, \boldsymbol{\phi}(x)\right). \tag{2.24}$$

In writing down a $2k$-leg operator—for example the two-leg operator $Q^{\alpha\beta}$—we can either use colour indices $\alpha > 1$, in which case the corresponding strand is forced to be finite and the correlator involves a Goldstone mode $\phi^{\alpha-1}$, or we can use $\alpha = 1$, in which case configurations in which the strand escapes to infinity dominate, and the leading behaviour of the correlator comes from taking $z^1 \sim 1$. For example $Q^{12} \sim \bar{\phi}^1$, while $Q^{23} \sim \phi^1\bar{\phi}^2$. Since in three dimensions the correlator of two Goldstone modes scales like $1/r$, the corresponding two-point functions are

$$\left\langle Q^{12}(x)Q^{21}(y)\right\rangle \sim |x - y|^{-1}, \quad \left\langle Q^{23}(x)Q^{32}(y)\right\rangle \sim |x - y|^{-2}. \tag{2.25}$$

That on the left is dominated by configurations in which a single infinite strand passes through both x and y, so that they are joined by a single finite strand, while that on the right gives the probability that x and y are joined by a single finite loop, i.e. by two finite strands.

For a system with only closed loops (e.g. with periodic boundary conditions) a similar separation into long and short strands is possible, using an infinitesimal magnetic field to fix the colours of the long strands.

We will discuss the extended phase in more detail in Sect. 2.9. Now we turn to behaviour at the phase transitions.

2.8 Critical Behaviour and First Order Transitions

We begin with a (very brief and incomplete) summary of Monte Carlo results. We then give an analytic picture for the RG flows in the CP^{n-1} model which sheds light on the numerical findings.

2.8.1 Summary of Numerical Results

The following numerical results were obtained by Pablo Serna, Miguel Ortuño, and Andres Somoza. A short account was published in Ref. [17], and a more detailed analysis in Ref. [18]. We take critical exponents for $n = 1$ from earlier work on the diamond lattice [3].

Numerically determined exponents at $n = 2$, which are $\nu = 0.708(5)$, $\gamma = 1.39(1)$ (here $\gamma = (2 - \eta)\nu$), are consistent with known $O(3)$ values [19], confirming the expected universality class of the transition. For $n > 2$, mean field theory predicts a first order transition, as a result of the cubic term in $\mathcal{L}_{\text{soft}}$; simulations suggest that the transition may instead remain continuous at $n = 3$, becoming first order only at larger n. Exponents for $n = 3$ were estimated as $\nu = 0.536(13)$, $\gamma = 0.97(2)$; see Ref. [18] for further details. Recent results for a (2+1)-dimensional bilayer $SU(3)$ magnet, which should be in the same universality class, are also consistent with a continuous transition [20].

For $n = 1$, the transition is continuous (with $\nu = 0.9985(15)$, $d_f = 2.534(9)$) as was first demonstrated for the loop model on the diamond lattice [3]. A continuous transition at $n = 1$ is in agreement with naive expectations: a first-order transition would require that the extended and short-loop phases coexisted at p_c, and that interfaces between them cost a nonzero surface tension; this seems to be ruled out by the independence of the nodes.

A convenient quantity for numerical analysis, which we will require again in Chap. 5, is the 'spanning number'. Consider first of all a system with open boundary conditions for the x direction, so that curves can end on the two faces at $x = 0$ and $x = L$. Then the spanning number n_s is the number of curves which propagate from $x = 0$ to $x = L$. (An equal number propagate in the reverse direction.) This quantity may be related to the free energy with twisted boundary conditions, as we will discuss in detail in Chap. 5, and its mean value $\langle n_s \rangle$ is equal to the spin stiffness in the CP^{n-1} model. In the localisation mapping, which exists for loop fugacity $n = 1$, $\langle n_s \rangle$ is the conductance [1–4].

In the extended phase, $\langle n_s \rangle \sim L$, while $\langle n_s \rangle$ decays exponentially with L in the short loop phase. In the vicinity of a critical point it has the finite-size scaling form

$$\langle n_s \rangle \simeq f\left(L^{1/\nu}(p - p_c)\right). \tag{2.26}$$

If the spanning number is plotted as a function of p, curves for different L are therefore expected to cross at p_c if the transition is continuous.

For computational convenience, we consider systems with periodic boundary conditions, and define n_s by slicing open the system; this is not expected to change the qualitative behaviour described above. Figure 2.8 shows crossings in $\langle n_s \rangle$ for $n = 2$ and $n = 3$ on the K lattice, for system sizes up to $L = 100$. This data leads to estimates for ν, either directly from the slopes of the curves at the crossings or from a scaling collapse using (2.26) plus corrections to scaling. (We present data here only

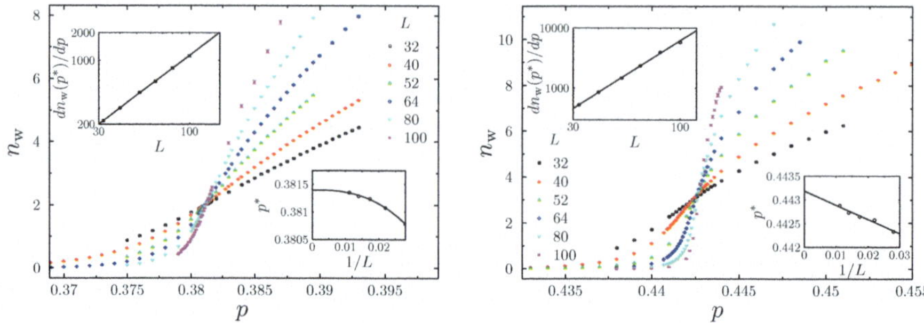

Fig. 2.8 *Left* mean spanning number at $n = 2$. *Main panel* number of spanning curves $n_w(p)$ plotted as a function of p for different system sizes L. *Upper inset* $dn_w(p)/dp$ at crossing points p^* versus L on log-log scales. *Lower inset* p^* versus L. *Right* same for $n = 3$

for the K lattice, but critical behaviour for $n = 1, 2, 3$ is similar on the L and K lattices, and is also consistent with simulations for $n = 1$ on the diamond lattice.)

We are not restricted to considering the mean $\langle n_s \rangle$; for example the probability P_k that n_s is equal to k has a similar finite size scaling form. This leads to the useful result that P_k is a universal, L-independent function of n_s near a critical point. Plotting (say) P_1 against $\langle n_s \rangle$ is thus a test for criticality, as the data should collapse without the need for any fitting parameters. Such plots are shown for $n = 2, 3, 4$ on the K lattice in Fig. 2.9 (left panel).

Data for $n = 2$ and $n = 3$ are compatible with continuous transitions, but stronger deviations from finite size scaling are seen at $n = 3$: we cannot say for certain whether this is a result simply of larger finite size effects, or whether the transition at $n = 3$ is first order, with an extremely large correlation length. Scaling collapse fails at $n = 4$, indicating a weakly first order transition; this is shown even more clearly in the double-peaked form of the probability distribution for N_p (Fig. 2.9, right). On theoretical grounds we expect that the transition at $n = 4$ on the L lattice is also first order, but the situation is complicated by its extreme proximity to $p = 1/2$,

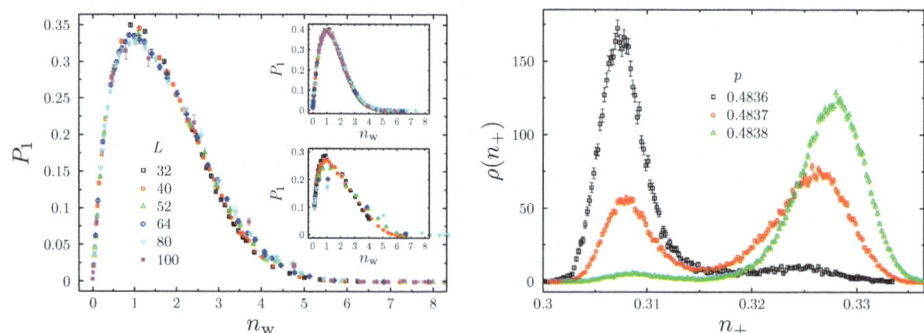

Fig. 2.9 *Left* P_1 plotted against $\langle n_s \rangle$ for $n = 2$ (*upper inset*), $n = 3$ (*main panel*) and $n = 4$ (*lower inset*). *Right* probability distributions for $N_p/(N_p + N_{1-p})$ on the K lattice at $n = 4$

where additional phenomena come into play (Chap. 3). Transitions at larger n are more strongly first order.

The RG picture of the following section implies that there is a universal n_c above which the transition in the CP^{n-1} model becomes first order. We may attempt a very crude estimate of n_c by extrapolating in n to the point at which the latent heat vanishes, using the data for the K lattice at $n \geq 4$. Naive polynomial fits give $n_c \gtrsim 3.3$. However, from the discussion below we expect the latent heat to vanish exponentially, $\sim A\exp(-B/\sqrt{n-n_c})$. Using the data for $n = 4, 5, 6$ to fix these constants gives $n_c = 3.0$. This estimate is not expected to be precise, since the exponential form is valid only asymptotically close to n_c, but it does indicate that n_c is likely to be fairly close to three.

2.8.2 The CP^{n-1} Model Near $n = 2$ and $d = 4$

At $n = 2$, when the cubic term $\mathrm{tr}\ Q^3$ in $\mathcal{L}_{\mathrm{soft}}$ vanishes, the upper critical dimension of the CP^{n-1} model is four rather than six.[2] This allows a double expansion in

$$\Delta = n - 2 \qquad \text{and} \qquad \epsilon = 4 - d. \tag{2.27}$$

This idea has been discussed previously for the Q-state Potts model, where the expansion is about the Ising limit [21]. The conclusions below for the CP^{n-1} model are qualitatively identical. In particular, a universal n_c appears, which is greater than the mean field value two when $d < 4$. This helps us to make sense of the numerical results of the previous section.

To begin with, recall the Wilson-Fisher [22] RG equations for the $O(3)$ (or CP^1) Lagrangian (2.13). To lowest nontrivial order these are (after rescaling $u \to u/22$)

$$\frac{\mathrm{d}u}{\mathrm{d}\ln L} = \epsilon u - u^2, \qquad \frac{\mathrm{d}t}{\mathrm{d}\ln L} = \left(2 - \frac{5}{11}u\right)t. \tag{2.28}$$

Setting $t = 0$, there are fixed points at $u = 0$ and $u = \epsilon$. For $\epsilon > 0$, the latter is stable in the u direction, and describes the critical CP^1 model, while the former is unstable in the u direction and describes the tricritical point.

We now consider a formal expansion of these equations in Δ (compare the approach to the $2+\epsilon$ dimensional $O(n)$ model in Ref. [23]), deferring the field-theoretic interpretation until Sect. 2.8.3. The leading contribution is a modification to the RG equation for u:

$$\frac{\mathrm{d}u}{\mathrm{d}\ln L} = -a\Delta + \epsilon u - u^2. \tag{2.29}$$

[2] See however Sect. 2.8.4.

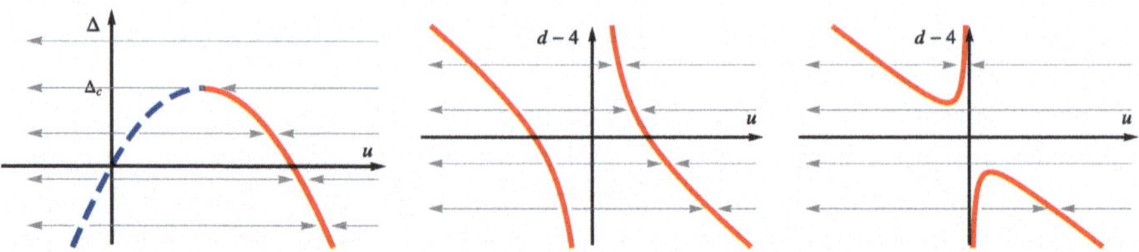

Fig. 2.10 *Left* RG fixed points and flows in the (u, Δ) plane for $d < 4$, showing critical (*red*) and tricritical (*blue, dashed*) fixed points merging at Δ_c. *Centre* RG fixed points and flows in the (u, d) plane for $n < 2$. *Right* the same for $n > 2$

The consistent scaling is to take Δ to be $O(\epsilon^2)$, and u to be $O(\epsilon)$ as at the Wilson Fisher fixed point. Here a is an undetermined universal constant, assumed positive in order to give sensible RG flows.

Figure 2.10 (left) shows the resulting RG fixed points in the (u, Δ) plane for fixed $\epsilon > 0$, i.e. slightly below four dimensions. As Δ is increased, the critical and tricritical fixed points approach each other, annihilating at a critical n_c given to this order by

$$n_c \simeq 2 + \frac{\epsilon^2}{4a}. \tag{2.30}$$

The thermal and leading irrelevant exponents at the critical point are

$$y_t \simeq 2 - \frac{5}{22}\left(\epsilon + \sqrt{\epsilon^2 - 4a\Delta}\right), \quad y_{\text{irr}} \simeq -\sqrt{\epsilon^2 - 4a\Delta}.$$

The anomalous dimension η is $O(\epsilon^2)$, as in the $O(N)$ model. Analogous formulas hold for the Potts model [21].

Precisely at n_c, the irrelevant exponent vanishes and there are logarithmic corrections to scaling. Above n_c, there are no fixed points: the RG flows go off to large negative u, which we interpret as a first order transition.

The strength of this first order transition decreases rapidly as n approaches n_c from above. Integrating Eq. 2.29 from a microscopic scale at which u is positive and of order one to the scale of the correlation length ξ, where u is negative and of order one, gives

$$\xi \sim \exp\left(\frac{2\pi}{\epsilon}\sqrt{\frac{n_c - 2}{n - n_c}}\right). \tag{2.31}$$

Similar forms hold for other quantities such as the latent heat [24]. Note that the asymptotic form $\xi \sim \exp\left(\text{const.}/\sqrt{n - n_c}\right)$ is more general than the lowest-order expansion we are considering here, depending only on the mechanism by which the critical point disappears at n_c [21, 24, 25].

In Fig. 2.10 we show the RG fixed points in the (d, u) plane for $n > 2$ and $n < 2$. Note that when $n < 2$ the fixed points below four dimensions are smoothly connected to those above. This is in agreement with our belief (see Sect. 4.4) that a $6-\epsilon$ expansion is possible in the model with $n = 1$. For $n > 2$, note the appearance of a branch of fixed points above four dimensions. These fixed points, which are at negative u, are not expected to correspond to genuine critical points, as a result of unboundedness of the fixed point potential (this phenomenon is present even in the $O(N)$ model [22]).

2.8.3 More Concrete Picture

Let us rewrite the soft spin Lagrangian for the CP^{n-1} model as follows:

$$\mathcal{L}_{\text{soft spin}} = \operatorname{tr}(\nabla Q)^2 + t \operatorname{tr} Q^2 + \frac{u}{22} \operatorname{tr} Q^4 + V(Q). \tag{2.32}$$

We have collected the operators which vanish at $n = 2$ into $V(Q)$. There is one of these at cubic and one at quartic order, but formally we allow for higher terms:

$$V(Q) = g_1 \operatorname{tr} Q^3 + g_2 \left(\operatorname{tr} Q^4 - \frac{1}{2} (\operatorname{tr} Q^2)^2 \right) + \cdots \tag{2.33}$$

The two operators shown above vanish when $n = 2$, by virtue of the tracelessness of Q (the second also vanishes at $n = 3$).

The RG equations for t and u must of course be independent of the g_i when $n = 2$, so any term which depends on the latter must have a coefficient proportional to Δ. A similar phenomenon also occurs in the Potts model [21]. To the order that we require, the RG equations for u and g_i are:

$$\frac{\mathrm{d}u}{\mathrm{d}\ln L} = \epsilon u - u^2 - \Delta f(g_1, g_2, \ldots), \qquad \frac{\mathrm{d}g_i}{\mathrm{d}\ln L} = \beta_i(g_1, g_2, \ldots). \tag{2.34}$$

(We have checked explicitly that the f term is of order Δ rather than higher order in Δ.) The cubic coupling g_1 is strongly relevant at the four-dimensional Gaussian fixed point, which obstructs a perturbative calculation of f and β_i. However to obtain Eq. 2.29 we need only assume that the g_i flow to fixed point values g_i^* under the RG equations (2.34), and expand around these with $g_i = g_i^* + \delta g_i$:

$$\frac{\mathrm{d}u}{\mathrm{d}\ln L} \simeq \epsilon u - u^2 - \Delta f(g_1^*, g_2^*, \ldots), \qquad \frac{\mathrm{d}\delta g_i}{\mathrm{d}\ln L} \simeq -b_{ij}\delta g_j. \tag{2.35}$$

The first of these equations yields Eq. 2.29, with $a = f(g_1^*, g_2^*, \ldots)$. The second yields subleading (order one) irrelevant exponents associated with the operators in $V(Q)$.

2.8.4 *Nonuniqueness of the Upper Critical Dimension at n = 2*

In order to analytically continue in n it is important to realise that it is the operators in $V(Q)$, and not their couplings, which vanish when $n = 2$. We may consider higher-dimensional versions of the loop models discussed here, and this observation leads to the conclusion that for $n = 2$ such models have two distinct upper critical dimensions.

For correlators which can be written down using only two spin indices, we do not need a replica limit (or SUSY). We set $n = 2$ directly, giving the $O(3)$ model with upper critical dimension four. These correlators will thus have Gaussian behaviour at the critical point for $d \geq 4$. However for correlators which require more than two indices, we are forced either to analytically continue in n or to use SUSY. In either case, the cubic term reappears (in the SUSY formulation, i.e. the soft-spin $\mathrm{CP}^{n+k-1|k}$ model, it is str Q^3, where str is the supertrace), leading to a non-Gaussian theory below six dimensions. These correlators are thus expected to have nontrivial behaviour for $d < 6$. This could be tested numerically by a simulation in four dimensions. (An appropriate ensemble of loops could be generated as a byproduct of the simulation of an $SU(2)$ magnet in $3+1$ dimensions.)

2.9 Joint Length Distribution in the Extended Phase

In the extended phase, a finite fraction of the links lie on ‘long’ loops, i.e. those with length of order L^3. How is the total length of these loops divided up between them? This is a natural question not only for the loop models considered here, but also for the worldlines appearing in quantum Monte Carlo and for the loops in a variety of other statistical physics problems (Sect. 2.11).

The answer is given by a one-parameter probability distribution known as the Poisson Dirichlet distribution [26] (here the parameter is n). This distribution appears in many situations which involve an infinite collection of random quantities, $\ell_1, \ell_2, \ldots$, whose sum is fixed. In particular, it has been rigorously proved to apply to the joint distribution of cycle lengths in natural ensembles of random permutations [27–29]—these can be viewed as loop ensembles with no spatial structure. It has been conjectured by Ueltschi and collaborators (with support from simulations) also to apply to loop ensembles with spatial structure, on the basis that when long loops proliferate spatial structure may become unimportant [30–32].

This conjecture for the loop length distribution has not however been related to field theory. Here we show how to derive it using the replica trick. The following derivation extends easily to other loop ensembles in an extended phase and to the random permutation problem.

Label the loops in a configuration $\mathcal{C}$ by $i = 1, \ldots |\mathcal{C}|$, and let ℓ_i denote the length of the ith loop. To begin with, consider the moments $\langle \sum_i \ell_i^m \rangle$ for integer $m > 1$. We obtain these by integrating the correlation function that gives the probability that m

links lie on the same loop (Eq. 2.17). Switching to a continuum notation, and with a being a nonuniversal operator normalisation constant,

$$\left\langle \sum_i \ell_i^m \right\rangle = \frac{n\,(m-1)!}{a^m} \int \mathrm{d}^3 x_1 \ldots \mathrm{d}^3 x_m \left\langle Q^{12}(x_1) \ldots Q^{m1}(x_m) \right\rangle. \tag{2.36}$$

To see this, note that in the graphical expansion of the above m-point function, a given loop configuration gives a nonzero contribution only if all the points x_k lie on the same loop i. For a given configuration and a given loop i, the integrals over x_k then give $\ell_i^m/(m-1)!$, where the factor in the denominator arises because the x_k must appear in a prescribed order around the loop.

In the ordered phase, the dominant contribution to the m-point function in (2.36) comes from the zero mode of $Q(x)$. The setting we are considering here, with only closed loops, corresponds to boundary conditions that preserve global $SU(n)$ symmetry, so we must average over the direction of symmetry breaking. Writing $Q(x) \simeq Q_0\,(\mathbf{z}\mathbf{z}^\dagger - 1)$, this is equivalent to averaging $\mathbf{z}$ over the sphere $\mathbf{z}^\dagger \mathbf{z} = n$. Such averages are given by

$$\mathrm{Tr}\; z^{\alpha_1} \ldots z^{\alpha_q} \bar{z}^{\beta_1} \ldots \bar{z}^{\beta_q} = \frac{n^{q-1} n!}{(n+q-1)!} \left[z^{\alpha_1} \ldots z^{\alpha_q} \bar{z}^{\beta_1} \ldots \bar{z}^{\beta_q} \right]_{\mathrm{G}}, \tag{2.37}$$

where $[\ldots]_{\mathrm{G}}$ is a Gaussian average evaluated using Wick's theorem and $[z^\alpha \bar{z}^\beta]_{\mathrm{G}} = \delta^{\alpha\beta}$. From this we have

$$\left\langle Q^{12}(x_1) \ldots Q^{m1}(x_m) \right\rangle \simeq Q_0^m \, \mathrm{Tr} \left(|z^1|^2 \ldots |z^m|^2 \right) = Q_0^m \frac{n^{m-1} n!}{(n+m-1)!}.$$

Combining this with the previous formula and defining $\mathcal{L}_{\mathrm{long}} = Q_0 L^3 n/a$,

$$\left\langle \sum_i \ell_i^m \right\rangle = \frac{n!(m-1)!}{(n+m-1)!} \mathcal{L}_{\mathrm{long}}^m. \tag{2.38}$$

Above, m is an integer greater than one (in order for the correlators to make sense). As a result, only long loops[3] contribute to the sum at leading order in L. We expect that we can analytically continue to $m = 1$ if we take the sum to run only over long loops, which we denote by a prime: $\langle \sum_i' \ell_i \rangle = \mathcal{L}_{\mathrm{long}}$. We see that $\mathcal{L}_{\mathrm{long}}$ is the total length of long loops. This can be confirmed by the arguments of Sect. 2.7 relating the order parameter to the fraction of links on long loops. We can also confirm using (2.41) below that the fraction of links on long loops is non-fluctuating in the thermodynamic limit.

[3] Picking any $u < 3$, only loops longer than L^u contribute at leading order. The fraction of links on long loops is not sensitive to precisely where (on a logarithmic scale) we put the division between long and short loops, since in the thermodynamic limit the total length is split between loops whose length is of order L^3 and those whose length is of order L^0.

In order to fix the full probability distribution we must calculate more general moments,

$$C(m_1, \ldots, m_q) \equiv \mathcal{L}_{\text{long}}^{-\sum_k m_k} \left\langle \sum\nolimits'_{i_1,\ldots,i_q} \ell_{i_1}^{m_1} \ldots \ell_{i_q}^{m_q} \right\rangle, \tag{2.39}$$

but these follow by an immediate extension of the above reasoning. We integrate an $(m_1 + \ldots + m_q)$-point correlation function with arguments

$$x_1^{(1)}, \ldots, x_{m_1}^{(1)} \,;\; \ldots \;;\, x_1^{(q)}, \ldots, x_{m_q}^{(q)}, \tag{2.40}$$

in which the coordinates $x_1^{(k)}, \ldots x_{m_k}^{(k)}$ are forced by a product of two-leg operators to lie on the same loop. A different set of spin indices is used for each k, so $\sum_{k=1}^{q} m_k$ distinct indices are used in total. We find

$$C(m_1, \ldots, m_q) = \frac{n^q\, \Gamma(n)\, \Gamma(m_1) \cdots \Gamma(m_q)}{\Gamma\left(n + m_1 + \cdots + m_q\right)}. \tag{2.41}$$

Take the ℓ_i to be ordered in decreasing size, and define normalised loop lengths by $\zeta_i = \ell_i / \mathcal{L}_{\text{long}}$ (so that $\sum'_i \zeta_i = 1$). The above formula shows that in the limit of large system size the probability distribution for the ζ_i is the Poisson-Dirichlet distribution, denoted PD(θ), with $\theta = n$.

This distribution may be defined as the limit of a Dirichlet distribution for N variables as $N \to \infty$. The Dirichlet distribution is given in terms of θ by

$$P_{\text{Dirichlet}}(\zeta_1, \ldots, \zeta_N) = \frac{\Gamma(\theta)}{\Gamma(\theta/N)^N} (\zeta_1 \zeta_2 \ldots \zeta_N)^{\theta/N - 1}\, \delta\Big(\sum\nolimits_i \zeta_i - 1\Big). \tag{2.42}$$

Calculating moments of (2.42) and taking the $N \to \infty$ limit reproduces the above formula for $C(m_1, \ldots, m_q)$.

Note that to obtain the full distribution we had to use an arbitrarily large number of colour indices. Translating the derivation into the SUSY language, this means we must consider the $\mathrm{CP}^{n+k-1|k}$ model for arbitrarily large k.

2.9.1 Length Distribution for a Single Loop

The above formulas allow us to extend our previous formula for the probability distribution $P_{\text{link}}(\ell)$ of the loop through a given link to the regime where ℓ is comparable with the system volume:

$$P_{\text{link}}(\ell) \simeq \begin{cases} c\, \ell^{-3/2} & (1 \ll \ell \ll L^2) \\ \theta\, \mathcal{L}_{\text{tot}}^{-1} \left[1 - \ell/\mathcal{L}_{\text{long}}\right]^{\theta - 1} & (L^2 \ll \ell \leq \mathcal{L}_{\text{long}}). \end{cases} \tag{2.43}$$

Recall that $\theta = n$. Here $\mathcal{L}_{\text{tot}}$ is the total loop length, given in our lattice models by $\mathcal{L}_{\text{tot}} = 3L^3/2$. In addition the above expressions involve two nonuniversal constants: c, and the fraction of links that lie on long loops, $f = \mathcal{L}_{\text{long}}/\mathcal{L}_{\text{tot}}$.

The above formula for the regime $\ell \gg L^2$ is obtained from the marginal distribution for a single ζ coming from Eq. 2.42:

$$P(\ell) = \frac{\ell \lim_{N\to\infty} N\, P_{\text{marg}}(\ell/\mathcal{L}_{\text{long}})}{\mathcal{L}_{\text{tot}} \times \mathcal{L}_{\text{long}}}, \quad P_{\text{marg}}(\zeta) = \Gamma(\theta)\frac{\zeta^{\theta/N-1}(1-\zeta)^{\theta-\theta/N-1}}{\Gamma(\theta/N)\,\Gamma([1-1/N]\theta)}.$$

The factor of ℓ in the first formula arises because a link chosen at random is more likely to lie on a long loop than a short one.

A notable feature of Eq. 2.43 is that $P(\ell)$ is non-monotonic when $\theta < 1$, developing a divergence at $\ell = \mathcal{L}_{\text{long}}$. This is perhaps not surprising, since the limit $n = 0$ describes the universality class of a single 'dense' polymer which fills the entire lattice: for this problem the entire probability distribution concentrates on $l = \mathcal{L}_{\text{tot}}$.

In the models discussed so far, where loops carry an orientation, we must consider $n < 1$ in order to see this divergence. However in the *unoriented* loop models discussed in the next section, the distribution PD(θ) appears with $\theta = n/2$—this follows by simple modifications of the above arguments—so the divergence can be seen at loop fugacity $n = 1$. (See also Sect. 2.11.) Further discussion of the loop length distribution, together with numerical results, may be found in Ref. [33].

2.10 Unoriented Loop Models

The loop models discussed so far are defined on oriented lattices, and the loops carry an orientation. We may define similar loop models on unoriented lattices, for example on the unoriented version of the L/K lattice. We then allow three distinct pairings at each node.

The lattice mapping (2.5) is easily adapted to such cases, leading to a field on real projective space, RP^{n-1}. Field theories similar to $\mathcal{L}_\sigma$ and $\mathcal{L}_{\text{soft}}$ apply, but with real-valued Q.

In three dimensions, the phenomenology—with transitions between short-loop and extended phases—is similar to that of the oriented loop models; preliminary numerical simulations for the case $n = 1$ (the RP^0 model) give $\nu \sim 0.9$ and $d_f \sim 2.54$ [34]. In two dimensions, however, the oriented and unoriented models show very different behaviour, as we will discuss in Chap. 5.

2.11 Loops in Frustrated Systems

In this section we conjecture field theories for 'fully packed' 3D loop models. These are similar to the completely packed models discussed above, but include an additional element of frustration which changes the continuum descriptions. The simplest example is the fully packed loop model on the cubic lattice. Unoccupied links are allowed, but every node is visited by precisely one loop:

$$Z_{\text{FPL}} = \sum_{\substack{\text{fully-packed} \\ \text{loop configs.}}} n^{\text{no. loops}}. \tag{2.44}$$

The limit $n = 0$ corresponds to a single dense polymer, or Hamiltonian walk, studied numerically in Ref. [35, 36]. (It would be interesting to use the field theory below to analyse these results.)

In analogy to various other constrained three-dimensional systems such as dimer models and spin ice, we expect the full-packing constraint to give rise to an emergent non-compact gauge field in the continuum [37, 38]. Such problems are said to be in a 'Coulomb' phase. We briefly recall the argument for such a field.

Let n_l be the occupation number of the link l, so that $n_l = 1$ if a loop passes through the link and $n_l = 0$ otherwise. Denoting the two sublattices of the cubic lattice by A and B, let $e_{l\mu}$ be the unit vector directed along link l from the site in A to the site in B ($\mu = 1, 2, 3$ is the spatial index). Now define the configuration-dependent vector field

$$B_{l\mu} = \left(n_l - \frac{1}{3}\right) e_{l\mu}. \tag{2.45}$$

The lattice divergence of this field, defined by a sum over the links adjacent to a given site i, vanishes as a result of full packing: $(\nabla_\mu B_\mu)_i = 0$. Assuming that B_μ provides an appropriate language for coarse-graining, and defining $B = \nabla \times A$ to take account of the divergence constraint, the simplest continuum Lagrangian for this field is

$$\mathcal{L}_{\text{Coulomb}} = \frac{\kappa}{2}(\nabla \times A)^2. \tag{2.46}$$

This Lagrangian is only part of the continuum description. For a complete one we must take into account all the conserved fluxes in the lattice model. $\mathcal{L}_{\text{Coulomb}}$ describes one type of conserved flux, which is carried by B. The loops are *not* flux lines for B. They do however carry another conserved flux, which is related to the fact that—after we have realised the loop fugacity via a sum over colours—the colour of a strand is conserved along its length. The interesting feature of this model is that while the two types of flux are made up of the same degrees of freedom microscopically, they decouple in the IR.

In order to see what the symmetries of the problem are, we write a lattice field theory analogous to Eq. 2.5:

$$Z_{\text{FPL}} = \text{Tr} \sum_{\{n_l\}} \delta(\nabla.B) \prod_{\langle ij \rangle} \left(\delta_{n_{ij},0} + \delta_{n_{ij},1} \mathbf{S}_i.\mathbf{S}_j\right). \tag{2.47}$$

The sum is over occupation numbers $n_l = 0, 1$, and the trace denotes the integral over $O(n)$ spins $\mathbf{S}$ living on the sites i (with the constraint $\mathbf{S}_i^2 = n$). A graphical expansion of the $\mathbf{S}$-dependent part of the Boltzmann weight generates the loop fugacity in the same manner as in Sect. 2.3.

In the absence of defects in the full-packing constraint, the Boltzmann weight for the spins in (2.47) has the $\mathbb{Z}_2$ gauge symmetry $\mathbf{S}_i \to \chi_i \mathbf{S}_i$, with $\chi_i = \pm 1$. The local gauge-invariant degree of freedom is therefore really a point on RP^{n-1} rather than on the sphere S^{n-1}. For brevity, we neglect this in the following discussion. This neglect is viable when $\mathbf{S}$ is ordered, i.e. in a phase with extended loops (see Chap. 5 for a more careful discussion of $\mathbb{Z}_2$ gauge symmetries in a 2D context). We expect (2.44) to yield an extended phase if n is smaller than some critical value; extended loops are of course guaranteed as $n \to 0$. We consider here only such a phase.

We also make the assumption that lattice symmetries are not spontaneously broken in the extended phase of (2.44). Such symmetry breaking would certainly be induced by adding sufficiently strong interactions favouring, for example, parallel loop segments. The resulting transitions are expected to correspond to Higgs transitions in which fluctuations of the gauge field become massive [39–42].

With these caveats, we conjecture the following continuum description for the extended phase:

$$\mathcal{L}_{\mathrm{FPL}} = \frac{K}{2}(\nabla \mathbf{S})^2 + \mathcal{L}_{\mathrm{Coulomb}} \qquad (\mathbf{S}^2 = 1). \tag{2.48}$$

We have rescaled the $O(n)$ spin in accordance with convention. The spin is ordered and exhibits Goldstone fluctuations; in this phase, symmetry-allowed interactions between A and $\mathbf{S}$ are irrelevant. Note that while A and $\mathbf{S}$ are independent in the IR, some operators—such as the one-leg operator, which involves a violation of the $\nabla . B$ constraint—combine both sectors. (Note also that at $n = 1$ the Goldstone modes associated with $\mathbf{S}$ vanish, so this field does not contribute to thermodynamic properties of the model; for this value of n it is only required for geometrical correlators.)

The simple idea above can be extended to other loop models. One is a classical loop model which appears in the study of the uniform resonating valence bond wavefunction for a spin-$1/2$ magnet on the cubic lattice [43]. (This wavefunction is the equal amplitude superposition of singlet coverings of the lattice, as described in Sect. 1.6.) Operator expectation values in this quantum state map to correlators in a classical loop model similar to (2.44), and a lattice partition function similar to 2.47 may be written. The naive continuum limit is again (2.48), with $n = 4$. This is consistent with the simulations of Ref. [43], which show the coexistence of long loops with Coulombic correlations.[4]

A class of two-flavour loop models may be approached in a similar manner. Classical spin ice is one special case of these models [44] and the four-colouring model on the pyrochore lattice is another [45]; in 2D they are well studied and can

[4] The $O(4)$ symmetry in this field theory is at first sight surprising, as it describes a wavefunction for an $SU(2)$ magnet. The symmetry is broken to $SU(2)$ if e.g. the amplitude of a singlet covering depends on the number of parallel singlets.

be understood using height model mappings [46, 47]. They involve configurations of two flavours of loops (call them + and − loops) on a four-coordinated lattice such as the diamond. Allowable configurations are those in which every node is visited by one + and one − loop. The two flavours can be assigned separate fugacities:

$$Z_{+-} = \sum_{\substack{\text{allowed} \\ \text{loop configs.}}} n_+^{+} n_-^{-}. \tag{2.49}$$

For these models we need two types of spin, one for each flavour, with n_+ and n_- components respectively. In some cases these models also have alternative descriptions which involve only gauge fields [45]; roughly speaking, the additional gauge fields are dual to the Goldstone modes in the description above.

For some of the models discussed in this section the loop length distribution has been determined numerically. The above descriptions and the arguments of Sect. 2.9 imply that the Poisson-Dirichlet distribution with $\theta = 1/2$ applies to the loops in spin ice studied in Ref. [44], that with $\theta = 1$ applies to those in the four-colouring model [45], and that with $\theta = 2$ applies to those in Ref. [43].

2.12 Conclusion

This chapter has described the relationship of a class of three-dimensional loop models to the CP^{n-1} model, and some simple consequences of this mapping for universal behaviour at phase transitions and in the extended phase. We also tackled some related issues such as the RG fixed point structure for the CP^{n-1} model.

When thinking about field theory in terms of random geometry—e.g. in terms of ensembles of loops—it should be borne in mind that a given critical field theory may be related to more than one universality class of geometrical model. The CP^{n-1} and RP^{n-1} models in particular are related to a remarkable variety of geometrical problems.

A simple example of this non-uniqueness is provided by CP^1. We have seen that this field theory is related to an ensemble of loops with fugacity $n = 2$. However, in its incarnation as the $O(3)$ model, it also has a well-known high temperature expansion that generates an ensemble of loops of fugacity three. These have different universal properties—for example, since the two-leg operator for these loops is no longer Q but instead a bilinear in Q, they have a different fractal dimension.

The difference between these two types of loops is essentially that the first kind are worldlines of $\mathbf{z}$, while the second are worldlines of the composite field Q. We might therefore ask whether worldlines of Q generate natural geometrical ensembles for other values of n. In two dimensions, this leads to an interesting relationship with Potts domain walls—this is described in Appendix A—as well as with branched polymers [48].

A feature of the CP^{n-1} models which we have not so far discussed is their relationship to gauge theories, in which the confinement of $\mathbf{z}$ into the composite field Q is effected by a fluctuating gauge field. This will become important when we discuss line defects in Chap. 4. It is also important for quantum magnets described by CP^{n-1} and related field theories [8]. In the next chapter we consider the connections between the loop models and such magnets.

References

1. I.A. Gruzberg, A.W.W. Ludwig, N. Read, Phys. Rev. Lett. **82**, 4524 (1999)
2. E.J. Beamond, J. Cardy, J.T. Chalker, Phys. Rev. B **65**, 214301 (2002)
3. M. Ortuño, A.M. Somoza, J.T. Chalker, Phys. Rev. Lett. **102**, 070603 (2009)
4. J. Cardy, in *50 Years of Anderson Localization*, ed. by E. Abrahams (World Scientific, Singapore, 2010)
5. A. Nahum, J.T. Chalker, Phys. Rev. E **85**, 031141 (2012)
6. M. Kamal, G. Murthy, Phys. Rev. Lett. **71**, 1911 (1993)
7. O.I. Motrunich, A. Vishwanath, Phys. Rev. B **70**, 075104 (2004)
8. T. Senthil, L. Balents, S. Sachdev, A. Vishwanath, M.P.A. Fisher, Phys. Rev. B **70**, 144407 (2004)
9. A.J. McKane, Phys. Lett. A **76**, 22 (1980)
10. G. Parisi, N. Sourlas, J. Phys. Lett. **41**, L403 (1980)
11. N. Read, H. Saleur, Nucl. Phys. B **613**, 409 (2001)
12. C. Candu, J.L. Jacobsen, N. Read, H. Saleur, J. Phys. A **43**, 142001 (2010)
13. K. Efetov, *Supersymmetry in Disorder and Chaos* (Cambridge University Press, Cambridge, 1997)
14. F. Haake, *Quantum Signatures of Chaos* (Springer, Berlin, 1991)
15. H. Saleur, B. Duplantier, Phys. Rev. Lett. **58**, 2325 (1987)
16. J. Kondev, C.L. Henley, Phys. Rev. Lett. **74**, 4580 (1995)
17. A. Nahum, J.T. Chalker, P. Serna, M. Ortuño, A.M. Somoza, Phys. Rev. Lett. **107**, 110601 (2011)
18. A. Nahum, J.T. Chalker, P. Serna, M. Ortuño, A.M. Somoza, Phys. Rev. B **88**, 134411 (2013)
19. M. Campostrini, M. Hasenbusch, A. Pelissetto, P. Rossi, E. Vicari, Phys. Rev. B **65**, 144520 (2002)
20. R.K. Kaul, Phys. Rev. B **85**, 180411 (2012)
21. K.E. Newman, E.K. Riedel, S. Muto, Phys. Rev. B. **29**, 302 (1984)
22. K.G. Wilson, M.E. Fisher, Phys. Rev. Lett. **28**, 240 (1972)
23. J.L. Cardy, H.W. Hamber, Phys. Rev. Lett. **45**, 499 (1980)
24. J.L. Cardy, M. Nauenberg, D.J. Scalapino, Phys. Rev. B **22**, 2560 (1980)
25. R.J. Baxter, J. Phys. C **6**, L 445 (1973)
26. J.F.C. Kingman, J. Roy. Statist. Soc. B **37**, 1 (1975)
27. O. Schramm, Israel J. Math. **147**, 221 (2005)
28. D. Ueltschi, J. Math. Phys. **54**, 083301 (2013)
29. C. Goldschmidt, D. Ueltschi, P. Windridge, Contemp. Math. **552**, 177 (2011)
30. S. Grosskinsky, A.A. Lovisolo, D. Ueltschi, J. Stat. Phys. **146**, 1105 (2011)
31. D. Ueltschi, J. Math. Phys. **54**, 083301 (2013)
32. C. Goldschmidt, D. Ueltschi, P. Windridge, Contemp. Math. **552**, 177 (2011)
33. A. Nahum, J.T. Chalker, P. Serna, M. Ortuño, A.M. Somoza, Phys. Rev. Lett. **111**, 100601 (2013)
34. P. Serna, M. Ortuño, A.M. Somoza, A. Nahum, J.T. Chalker, Critical Behaviour in 3D Unoriented Loop Models (in preparation)

35. J. Jacobsen, Phys. Rev. Lett. **100**, 118102 (2008)
36. J. Jacobsen, Phys. Rev. E **82**, 051802 (2010)
37. D.A. Huse, W. Krauth, R. Moessner, S.L. Sondhi, Phys. Rev. Lett. **91**, 167004 (2003)
38. C.L. Henley, Annu. Rev. Condens. Matter Phys. **1**, 179 (2010)
39. S. Powell, J.T. Chalker, Phys. Rev. Lett. **101**, 155702 (2008)
40. D. Charrier, F. Alet, P. Pujol, Phys. Rev. Lett. **101**, 167205 (2008)
41. S. Powell, J.T. Chalker, Phys. Rev. B **80**, 134413 (2009)
42. G. Chen, J. Gukelberger, S. Trebst, F. Alet, L. Balents, Phys. Rev. B **80**, 045112 (2009)
43. A. Fabricio Albuquerque, F. Alet, R. Moessner, Phys. Rev. Lett. **109**, 147204 (2012)
44. L.D.C. Jaubert, M. Haque, R. Moessner, Phys. Rev. Lett. **107**, 177202 (2011)
45. V. Khemani, R. Moessner, S.A. Parameswaran, S.L. Sondhi, Phys. Rev. B **86**, 054411 (2012)
46. J.L. Jacobsen, J. Kondev, J. Stat. Phys **96**, 21 (1999)
47. J.L. Jacobsen, J. Kondev, Nucl. Phys. B **532**, 635 (1998)
48. J. Cardy, J. Phys. A: Math. Gen. **34**, L665 (2001)

Chapter 3
Topological Terms, Quantum Magnets and Deconfined Criticality

3.1 Introduction

We saw in the previous chapter that the completely-packed loop models are described by lattice magnets in which the Boltzmann weight is generically complex. At the critical points discussed so far, the corresponding coarse-grained action was nevertheless both real and of the form that would be naively expected on the grounds of symmetry. But, depending on the choice of lattice and interactions, it may be necessary to consider the contribution of imaginary terms to the Boltzmann weight more carefully.

On the 2D L lattice (Fig. 2.1), they lead to a topological θ term in the continuum action, consistent with the result from the transfer matrix mapping to a (1+1)D spin chain [1, 2]. For the 3D L lattice, they cause hedgehog defects in the CP^{n-1} spin configuration to acquire complex fugacities which vary in space. This can be seen either by directly coarse-graining the action for the loop model, or by mapping the loop model to a $(2+1)$D magnet on the square lattice and using known results for problems of this kind [3–5]. The nontrivial hedgehog fugacities can result in a phase cancellation effect that modifies the critical behaviour, in line with the scenario known as deconfined criticality [6, 7]. To see deconfined critical behaviour in the loop model, we must introduce an additional coupling that induces a transition without explicitly spoiling the lattice symmetries (i.e. while remaining at $p = 1/2$).

The purposes of this short chapter are to review some background material on the CP^{n-1} model (in both two and three dimensions) that will be needed later; to sketch how the abovementioned lacuna in our previous discussion of coarse-graining the loop models is filled in; and to point out a connection between the loop models and quantum magnets on the square lattice (which makes the loop models a useful platform for studying quantum criticality). We begin by discussing the role of spin configurations with nontrivial topology in the CP^{n-1} model, the formulation of this field theory as a gauge theory, and the related 'noncompact' CP^{n-1} model (NCCP^{n-1}) that has been argued to appear in 3D when hedgehogs are suppressed [6–8]. We then indicate how hedgehogs on the L lattice acquire complex fugacities. Next, we

A. Nahum, *Critical Phenomena in Loop Models*, Springer Theses,
DOI 10.1007/978-3-319-06407-9_3

sketch the transfer matrix mapping to $SU(n)$ magnets on the square lattice. Finally, we discuss how to modify the interactions for the L lattice loop model in order to see the deconfined transition. For brevity, much of our discussion will be heuristic; additionally, we will not include here numerical results for the deconfined transition in the loop model, which will appear in Ref. [9].

3.2 Topology and Gauge Fields in the CP^{n-1} Model

The fundamental group of the manifold CP^{n-1} is trivial, indicating that the configuration $Q(x)$ cannot display vortex defects. However the second homotopy group $\pi_2(\mathrm{CP}^{n-1})$ is equal to $\mathbb{Z}$ for any $n > 1$. As a result, $Q(x)$ can sustain smooth skyrmion textures in two dimensions [10, 11], and pointlike hedgehog defects in three [12]. While we are concerned at present with 3D, we also review the 2D case [1, 10, 11, 13] for future reference.

3.2.1 *Two Dimensions: Skyrmions and the θ Term*

Skyrmions lead to the possibility of a θ term in the 2D Lagrangian:

$$\mathcal{L}_\sigma = \frac{K}{4} \operatorname{tr} (\nabla Q)^2 + \frac{\theta}{2\pi} \epsilon_{\mu\nu} \operatorname{tr} Q \nabla_\mu Q \nabla_\nu Q. \tag{3.1}$$

We have returned to the conventional normalisation $|\mathbf{z}|^2 = 1$, $Q = \mathbf{z}\mathbf{z}^\dagger - \mathbb{1}/n$, which we will retain from now on when we write Lagrangians for sigma models.

The new expression is purely imaginary, and is proportional to the topological density: integrated over a closed manifold, it gives $i\theta$ multiplied by the (signed) number of skyrmions. For $n = 2$, setting $S = \operatorname{tr} \sigma Q$ gives the conventional form of the $O(3)$ sigma model with θ term. A pedagogical discussion of this case may be found in Ref. [12].

As is clear from the skyrmion picture, bulk properties of the sigma model depend on θ only modulo 2π. A parity transformation changes the sign of θ, so only the values $\theta = 0$ and $\theta = \pi \pmod{2\pi}$ preserve parity.

If θ is *not* equal to $\pi \pmod{2\pi}$, the model is massive (at least for $n > 0$ [2, 14]) and flows under RG to $\theta = 0 \bmod 2\pi$, $K = 0$. However if $\theta = \pi$, and if in addition $n \leq 2$ and the bare stiffness K is sufficiently large, the model flows to a nontrivial fixed point. For $n = 2$, this is the fixed point governing the spin-1/2 antiferromagnetic chain, in which the θ term arises from the imaginary Berry phases that constitute the action for a single spin [12]. The relevant perturbation $(\theta - \pi)$ is induced by dimerising the couplings in the spin chain.

For general n, this fixed point governs the loop model on the 2D L lattice (Fig. 2.1) at its critical point $p = 1/2$ [1, 2], and other loop models in the dense phase (Sects. 1.2

and 4.5). On the L lattice, the perturbation $(\theta - \pi)$ corresponds to varying p away from 1/2: this favours (say) clockwise-turning over anticlockwise-turning nodes, and immediately induces a phase with short loops.[1]

(The fixed point discussed so far is that to which the sigma model with large K and $\theta = \pi$ flows; it is stable in the K direction for $n < 2$ and marginally stable for $n = 2$. For $n < 2$, appropriate lattice regularisations of the CP^{n-1} model will also display an unstable fixed point at smaller K, which corresponds to the dilute critical point—see Sect. 4.5 and Ref. [15]. This fixed point is also unstable in the $(\theta - \pi)$ direction.)

The equivalence between Eq. 3.1 and the formulation of the CP^{n-1} model given in Sect. 1.3.2 will appear in Sect. 3.2.3.

3.2.2 Three Dimensions: Hedgehogs

Any regularisation of the sigma model in three dimensions will determine a fugacity, or core action, for hedgehogs. Since they are pointlike defects, they are not counted by a term in the continuum action analogous to the θ term above; however, integrating the same topological density over a closed 2D surface gives the signed number of hedgehogs inside. Hedgehogs are of course irrelevant in the ordered phase, but theoretical and numerical work indicates that their presence or absence is crucial in determining the universal behaviour at the transition [6–8, 16–21]. In certain situations the coarse-grained hedgehog fugacity can vanish, and the appropriate field theory is then no longer that of $\mathcal{L}_{\text{soft}}$ (the soft-spin Lagrangian of Eq. 2.11) but the $NCCP^{n-1}$ model, which we will review in the next section [6–8].

Several mechanisms for the suppression of hedgehogs have been considered. Firstly, one may include an explicit no-hedgehog constraint in a classical lattice $O(3)$ model [8, 16]. Secondly, hedgehogs are RG-irrelevant at the $NCCP^{n-1}$ critical point for sufficiently large n [22, 23]; correspondingly, certain lattice regularisations of the CP^{n-1} model, in which hedgehogs are allowed microscopically, show a continuous $NCCP^{n-1}$ transition at large[2] n rather than the first order transition predicted by $\mathcal{L}_{\text{soft}}$. Other lattice regularisations of course give first order transitions at large n [10]. Finally, hedgehogs can be suppressed by the phase cancellation effect mentioned in the introduction [3–7]. More precisely, this cancellation forbids hedgehogs of unit topological charge, but allows for hedgehogs of higher charge (multiples of four in the case of interest here); these are believed to be irrelevant at the critical point, but are important in the disordered phase [3–5].

It will be useful to review the formulation of the CP^{n-1} model as a gauge theory before returning to these issues in the context of the loop model.

---

[1] This is true for $0 < n \leq 2$. The completely packed loop model at $n = 0$ (the dense polymer) is different since we cannot have short loops: $(\theta - \pi)$ is marginal there [2, 14].

[2] The precise value of n beyond which hedgehogs of unit topological charge become irrelevant at the $NCCP^{n-1}$ fixed point is not currently known, but is larger than $n = 10$ [24].

3.2.3 Formulation as a Gauge Theory

Using $Q = (\mathbf{z}\mathbf{z}^\dagger - 1/n)$, the sigma model Lagrangian $\mathcal{L}_\sigma = \frac{K}{4}\,\mathrm{tr}\,(\nabla Q)^2$ may be written

$$\mathcal{L}_\sigma = \frac{K}{2}\,|(\nabla - iA)\,\mathbf{z}|^2\,, \tag{3.2}$$

with a gauge field defined by

$$A = \frac{i}{2}\left((\nabla\mathbf{z}^\dagger)\mathbf{z} - \mathbf{z}^\dagger(\nabla\mathbf{z})\right). \tag{3.3}$$

In two dimensions, the field strength for A is $B = \epsilon_{\mu\nu}\nabla_\nu A_\nu = \frac{1}{i}\epsilon_{\mu\nu}\nabla_\mu(\mathbf{z}^\dagger\nabla_\nu\mathbf{z})$. This is proportional to the topological density in (3.1), so to include the θ term we add $i\theta B/2\pi$ to (3.2). In three dimensions, $B_\mu = \frac{1}{i}\epsilon_{\mu\nu\lambda}\nabla_\mu(\mathbf{z}^\dagger\nabla_\nu\mathbf{z})$.

Above, A is defined in terms of $\mathbf{z}$. But we may instead take A to be an independent fluctuating field in (3.2); performing the Gaussian integral over this field is equivalent to replacing it with its classical value, which is (3.3). So the CP^{n-1} model may be viewed as a $U(1)$ gauge theory, as in Sect. 1.3.2. This formulation also extends straightforwardly to the supersymmetric $\mathrm{CP}^{n+k-1|k}$ model, by replacing $\mathbf{z}$ with $\boldsymbol{\psi}$ [1]. The k-independence of the partition function for the SUSY model can be seen by using an auxiliary field to enforce the length constraint on $\boldsymbol{\psi}$: the integral over $\boldsymbol{\psi}$ is then Gaussian, and gives a determinant to the power of $k - (n + k)$.

The Lagrangian (3.2) looks unconventional, in that it lacks a kinetic term for A. This is not in itself important—for example we may imagine one appearing upon coarse-graining. But what is crucial, in three dimensions, is that A is a compact gauge field. In a lattice regularisation, this means that the gauge degree of freedom is a periodic variable [10]. In the continuum, it implies that Dirac monopole configurations in A are permitted [25, 26]. (We will discuss monopoles again in Chap. 4. See Refs. [25–27] for discussions of compact gauge theory.)

In three dimensions, the consequence of the compactness of the gauge field is confinement. This is the heuristic reason why a field theory with $\mathbf{z}$ coupled to a compact gauge field is equivalent to $\mathcal{L}_\sigma$ or to $\mathcal{L}_{\mathrm{soft}}$: since the charged degrees of freedom in $\mathbf{z}$ are confined, only the neutral ones in Q need be taken into account on large scales.

3.2.4 Noncompact CP^{n-1} Model

So far, the above is merely an alternative formulation for a familiar universality class (albeit one that will be useful to us in Chap. 4). Its utility for describing other kinds of critical behaviour arises from the correspondence between hedgehogs in Q and monopoles in A. This correspondence follows from the identification of the field strength with the topological density: the flux emanating from a sphere is equal to the number of monopoles enclosed (when contributions from singular Dirac strings are discounted).

As a result it is believed that critical behaviour in the models in which hedgehogs are suppressed is captured by a Lagrangian in which $\mathbf{z}$ is coupled to a non-compact gauge field [6–8]. (For the $SU(n)$ magnets, this conclusion is also supported by arguments in a dual language [28].) In a soft-spin formulation,

$$\mathcal{L}_{\mathrm{NCCP}^{n-1}} = |(\nabla - iA)\mathbf{z}|^2 + \kappa B^2 + \mu|\mathbf{z}|^2 + \lambda|\mathbf{z}|^4. \tag{3.4}$$

It should be noted that the ordered phase, where A is rendered massive by the Higgs mechanism, is insensitive to whether A is compact or noncompact; similarly, hedgehogs are RG irrelevant so their suppression is a moot point. On the other hand noncompactness changes the properties of the disordered phase, in which $\mathbf{z}$ is massive. In the NCCP^{n-1} model, there remains a massless 'photon' degree of freedom associated with the gauge field [8], while in the compact case A is rendered massive by monopole proliferation. However, for the $(2+1)D$ magnets on the square lattice, and the L lattice loop model at $p = 1/2$, we are not strictly dealing with the noncompact theory in which all monopoles are suppressed, but rather with a theory in which quadrupled monopoles are allowed. These are believed to be irrelevant at the critical point, but are relevant in the disordered phase, and remove the massless photon.

The NCCP^1 model arises in a different way as a description of 3D classical dimer models [29–33]. There, the gauge symmetry is non-compact even in the microscopic model, so the disordered phase does indeed have a massless gauge field. This is a Coulomb phase of the type discussed in Sect. 2.11. Instead it is the $SU(n)$ symmetry that is argued to be emergent at the transition (so it is the Goldstone modes in the ordered phase that are lost as a result of terms that break $SU(n)$ symmetry). In both contexts, magnets and dimer models, numerical results for the transition are perplexing, and show large corrections to scaling whose origin is still not understood.

3.3 Hedgehogs on the L Lattice at $p = 1/2$

In the previous chapter we were cavalier about coarse-graining the loop models, so let's return to the lattice partition function (2.5). Specifically, consider the L lattice at $p = 1/2$. Recall the form of the partition function Z (for convenience, we again rescale $\mathbf{z}$ so that $|\mathbf{z}|^2 = 1$):

$$\mathrm{Tr}\exp(-S) = \mathrm{Tr}\prod_{\text{nodes}}\left(\frac{1}{2}(\mathbf{z}_o^\dagger\mathbf{z}_i)(\mathbf{z}_{o'}^\dagger\mathbf{z}_{i'}) + \frac{1}{2}(\mathbf{z}_o^\dagger\mathbf{z}_{i'})(\mathbf{z}_{o'}^\dagger\mathbf{z}_i)\right). \tag{3.5}$$

To begin with, pretend that $\mathbf{z}$ is slowly varying. Each term in the above product is then close to one, and we may imagine obtaining a continuum sigma model action by a derivative expansion. In three dimensions,[3] the only term with two derivatives allowed by global, gauge and lattice symmetries is the kinetic term in $\mathcal{L}_\sigma$, whose bare coupling may be estimated rapidly at $p = 1/2$ by calculating the right hand side of (3.5) for a $\mathbf{z}$ configuration with a uniform twist, giving $K = 1/4$.

This derivative expansion can fail to capture the true action in two ways. Firstly, it fails in a trivial way if the gauge-invariant spin Q varies strongly at a node. This will of course be the case in the lattice model, and leads to an order one renormalisation of the bare stiffness even for those values of n for which the model is in the ordered phase.[4] Of course, we could control the derivative expansion by artificially suppressing fast variations of Q, for example by adding the gauge-invariant term $\lambda \sum_{\langle ll' \rangle} |\mathbf{z}_l^\dagger \mathbf{z}_{l'}|^2$ to the action with $\lambda \gg 1$.

Secondly, even if Q is slowly varying, the *phase* of $\mathbf{z}$ can vary rapidly. Nodes where Q is approximately constant but where this phase varies abruptly contribute imaginary terms to the action. For smooth configurations of trivial topology, these phases cancel. However, in the presence of hedgehog defects, nontrivial phases that are missed by the derivative expansion can remain.

This is because in a configuration with a hedgehog it is impossible to find a gauge in which $\mathbf{z}$ is everywhere slowly varying, even far from the hedgehog core. This follows from the fact that the expression for the topological density is a total derivative when written in terms of $\mathbf{z}$ (it is B_μ, given below Eq. 3.3). If $\mathbf{z}$ were continuous, integrating the topological density over a large sphere would give zero, which cannot be the case if a hedgehog is enclosed.

A simple calculation is required to determine the effect this has on the action for a hedgehog configuration. This is described in Appendix B. For specificity, we take the hedgehog to be centred on a site of C_1 or a site of C_2—for example at the centre of the cube in Fig. 2.4. These locations form a bcc lattice, with four sublattices. The result (which we might also guess from the mapping described in the next section and known results for quantum magnets [3–5]) is that the hedgehog fugacity is proportional either to $1, i, -1$ or $-i$, depending on which sublattice it sits on.[5]

We infer that the coarse-grained hedgehog fugacity vanishes on the L lattice at $p = 1/2$. It is important to note that this vanishing relies upon lattice symmetry, and is specific to the L lattice at $p = 1/2$—it does not hold for the critical points discussed in the previous chapter.

However, by adding an additional interaction to the model, we can drive a transition without breaking lattice symmetry by hand. Specifically, we introduce an

[3] On the 2D L lattice, coarse-graining also yields the θ term with $\theta = \pi$. A trick to obtain this coefficient is to consider a system with boundary and choose a gauge for $\mathbf{z}$ in which this term reduces to a sum over the boundary links.

[4] At $n = 1$, $p = 1/2$, the stiffness is approximately $K \simeq 0.4$ (calculated numerically from the ratio of the spanning number to system size).

[5] More precisely, only the relative phase between these locations is meaningful; for example for a finite system with a single hedgehog, the absolute phase depends on the choice of boundary conditions for $\mathbf{z}$. An antihedgehog (of opposite topological charge) has the opposite phase.

interaction between nearest neighbour nodes on the same sublattice (between nodes of like colour in Fig. 2.4). Each node constitutes a binary degree of freedom which we represent with an Ising-like variable, $\sigma = \pm 1$: the four short-loop configurations correspond to the four states in which σ is perfectly ordered. The interaction is ferromagnetic for the σs on each sublattice.

3.4 Transfer Matrices and Quantum Magnets

There is a correspondence between the 3D classical models of the last chapter and 2D quantum magnets. The latter provide a strong motivation for studying critical behaviour in the CP^{n-1} model and related field theories, so we now describe this correspondence. We first review a standard toy example, then state the qualitative features of the Hamiltonians corresponding to the 3D loop models.

The link colours provide a convenient basis for the transfer matrix, which acts between 'time' slices formed by planes of links. Consider first of all a single node of the type shown in Fig. 3.1. In component form this is $T_{\alpha\beta}^{\alpha'\beta'}$, where the upper indices are the colours of the links at time $\tau + \Delta\tau$, and the lower indices those at time τ. This matrix has two terms, corresponding to the two node configurations:

$$T_{\alpha\beta}^{\alpha'\beta'} = (1-p)\delta_{\alpha}^{\alpha'}\delta_{\beta}^{\beta'} + p\,\delta^{\alpha'\beta'}\delta_{\alpha\beta}. \tag{3.6}$$

The partition function $\operatorname{tr} T^N$ defines a trivial quasi-one-dimensional loop model with N nodes. The transfer matrix is also the imaginary time evolution operator for a two-site quantum problem with Hamiltonian H:

$$T = e^{-\Delta\tau H}. \tag{3.7}$$

Each site (we label them A and B) has an n-dimensional Hilbert space, spanned by kets $|\alpha\rangle_A$ and $|\alpha\rangle_B$ respectively. In terms of these,

$$T = (1-p)\mathbb{1} + p\sum_{\alpha\beta} |\alpha\rangle_A\,|\alpha\rangle_B\,\langle\beta|_A\,\langle\beta|_B = (1-p)\mathbb{1} + p\,n\,\mathcal{P}_{AB}. \tag{3.8}$$

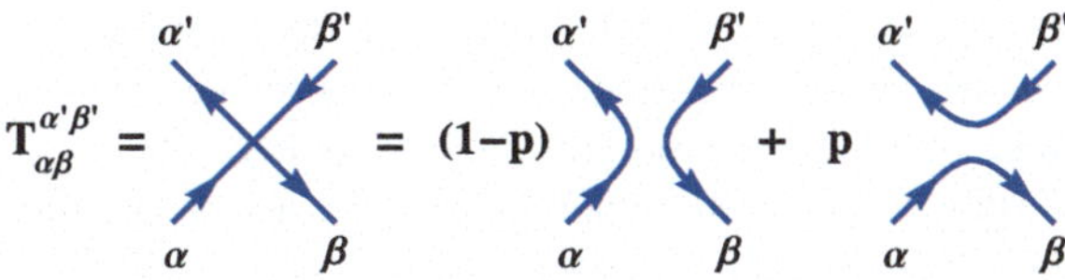

Fig. 3.1 Graphical representation of the transfer matrix for a single node. The indices $\alpha, \beta, \alpha', \beta'$ denote link colours. Imaginary time flows vertically, with the lower links being at time τ and the upper links at time $\tau + \Delta\tau$

$\mathcal{P}_{AB}$ is the projector onto the state $\frac{1}{\sqrt{n}}\sum_\alpha |\alpha\rangle_A |\alpha\rangle_B$, which is a singlet under $SU(n)$ transformations acting as

$$|\alpha\rangle_A \longrightarrow U_{\alpha\beta}\,|\beta\rangle_A\,, \quad |\alpha\rangle_B \longrightarrow U^*_{\alpha\beta}\,|\beta\rangle_B\,. \tag{3.9}$$

The degrees of freedom at A and B are $SU(n)$ spins, transforming in the fundamental and antifundamental representation respectively.

At $n = 2$ they are standard spin$-1/2$ degrees of freedom. In this case it is convenient to relabel the basis states as S_z eigenstates for spin operators $\mathbf{S}_A$ and $\mathbf{S}_B$,

$$\begin{aligned} |1\rangle_A &= |\uparrow\rangle_A\,, \quad |1\rangle_B = |\downarrow\rangle_B\,, \\ |2\rangle_A &= |\downarrow\rangle_A\,, \quad |2\rangle_B = -\,|\uparrow\rangle_B\,. \end{aligned} \tag{3.10}$$

Both spins then transform in the same representation (this is possible at $n = 2$ because the fundamental representation of $SU(2)$ is pseudoreal). The singlet takes the usual form $\frac{1}{\sqrt{2}}[|\uparrow\rangle_A |\downarrow\rangle_B - |\downarrow\rangle_A |\uparrow\rangle_B]$, and the projector onto it is $\mathcal{P}_{AB} = 1/4 - \mathbf{S}_A.\mathbf{S}_B$.

For the single node we can easily extract the form of the Hamiltonian. Dropping an additive constant, it is $H = J\,\mathcal{P}_{AB}$ with $J = \Delta\tau^{-1}\ln[1 + pn/(1-p)]$. For 2D or 3D loop models, the Hamiltonian will not generally take a simple explicit form. It is possible to obtain one by taking a continuum limit in imaginary time—i.e. by making the node weights anisotropic in such a way that the transfer matrix becomes close to the identity. This procedure is standard for the loop model on the two-dimensional L lattice. The resulting Hamiltonian describes a nearest-neighbour spin chain, and the node parameter in the loop model controls staggering in the strength of the exchange [2]: the critical point, at which $\theta = \pi$ in the continuum description (3.1), is where the staggering vanishes.

For the 3D models, we take the view that the precise form of H is less important than the degrees of freedom, symmetries and phase structure of H. For both the L and K lattice we take imaginary time to run parallel to the z axis. The links piercing the boundary of a time slice (which is of thickness 4 for the L lattice and 2 for the K) then form a square lattice with lattice spacing $\sqrt{2}$, as shown in Fig. 3.2. One sublattice consists of upgoing and the other of downgoing links; as in the single-node example, this leads to an $SU(n)$ magnet with spins in the fundamental representation on one sublattice and in the antifundamental representation on the other. The transfer matrix T for a given lattice is a sum over configurations within the timeslice, in analogy to Eq. 3.6. In the absence of the inter-node interaction described at the end of the last section, the phase structure of the models is like that of nearest-neighbour $SU(n)$ magnets with dimerization in the strength of the exchange—$H = \sum_{\langle ij\rangle} J_{ij}\mathcal{P}_{ij}$ with J_{ij} stronger on certain bonds—although of course the actual Hamiltonian (given by the logarithm of T) does not take this simple form.

The extended phase in the loop models corresponds to the Néel phase, while short-loop phases correspond to dimerised phases. The pattern of dimerisation in a short-loop phase can be seen from the representative configuration in which all loops

Fig. 3.2 The *square* lattice for the associated quantum problem is formed by the links in a time slice (This shows the K lattice at $p = 0$: this short loop phase corresponds to a staggered dimer state.)

have the minimal length of six. These loops connect the links within a timeslice in pairs; in the quantum problem singlets form on the paired sites. For the K lattice there is only a single packing of minimal-length loops, which corresponds to a staggered packing of singlets.

The L lattice is more interesting. There are four packings of minimal-length loops, and they map to the four columnar packings of singlets (valence bonds) on the square lattice. When $p = 1/2$ these packings are related by lattice symmetry, which is broken when $p \neq 1/2$: varying p away from $1/2$ corresponds to dimerising the couplings in H so as to favour one of the columnar packings. The transitions obtained by varying the dimerisation in this way are known to be described by the compact CP^{n-1} model [21, 34]. They are transitions which involve only one order parameter, namely Q.

As has been much discussed recently, we may also consider transitions from the Néel state to a valence bond solid (VBS), in which the symmetry between the four dimerised states is broken spontaneously rather than by the Hamiltonian. In order to obtain such a transition, the nearest-neighbour coupling can be supplemented by a term which favours parallel pairs of singlets, giving the so-called J–Q model [17]. In this case there is an additional order parameter to consider. This is nonzero in the VBS phase and distinguishes between the four dimerised states. It may be viewed as a planar spin ϕ with $\mathbb{Z}_4$ anisotropy.

While a direct transition between the two phases would be first order according to a naive application of Landau theory (since fine-tuning would be required in order for the masses of ϕ and Q to vanish simultaneously) it has been argued that a continuous one described by the NCCP^{n-1} model can occur instead. The NCCP^{n-1} model may be argued for either as in Sect. 3.2.4 [6, 7], or by an alternative argument in which the ϕ configuration is considered [28]. The key point is that vortices in ϕ carry a single unpaired spin. Combining this observation with standard duality arguments for the XY model (which will appear again in the following chapter) leads again to the NCCP^{n-1} model. There is a correspondence between the operator which inserts a monopole in the NCCP^{n-1} model and the order parameter ϕ; the presence of quadrupled monopoles corresponds to $\mathbb{Z}_4$ anisotropy in the microscopic distribution of ϕ, and their irrelevance at the NCCP^{n-1} critical point leads to emergent $U(1)$ symmetry in the critical probability distribution for ϕ.

As mentioned in Sect. 3.3, we can see the deconfined transition in the L lattice loop model by adding an interaction between nodes, and the order parameter ϕ is then defined in terms of the node configurations. Numerical results, on systems up to $L = 512$ at $n = 2$, indeed show a direct transition between the Néel (extended) and VBS (localised) phases [9]. There are none of the usual signs of first order behaviour, and the characteristic features of the deconfined critical point, including the emergent $U(1)$ symmetry for ϕ and a large anomalous dimension for Q (or S) are seen. However, in common with numerical work on the J–Q model [18–20, 35], we see strong violations of finite size scaling. These are manifested particularly drastically in the failure of P_1 to collapse when plotted against n_s (see Sect. 2.8 for definitions). The origin of these scaling violations is not understood, and they call for a deeper understanding of the RG flows for the NCCP^{n-1} model.

3.5 Conclusion

In view of the above mapping, we may regard the loop models as laboratories for the numerical investigation of critical behaviour in quantum magnets. This correspondence is of course only useful for studying universal behaviour, since the transfer matrix does not give a simple or explicitly-known quantum Hamiltonian. It is also not the only way to use loops to simulate the magnets—modern worldline-based algorithms deal with loop ensembles that are akin to those considered here [34]. However, the loop models have some advantages. They may be regarded as models that are designed so that the transfer matrix, rather than the Hamiltonian, is simple. This leads to simplifications in numerics, for example as a result of the fact that the space and imaginary time dimensions are treated on the same footing microscopically—this eliminates uncertainty about the dynamical critical exponent, and fixes one (usually unknown) constant, namely the theory's characteristic velocity. The loop models also suggest new observables, such as the spanning number discussed in Chap. 2—this is similar to, but distinct from, the winding numbers used in quantum Monte Carlo calculations.

References

1. N. Read, H. Saleur, Nucl. Phys. B **613**, 409 (2001)
2. C. Candu, J.L. Jacobsen, N. Read, H. Saleur, J. Phys. A **43**, 142001 (2010)
3. F.D.M. Haldane, Phys. Rev. Lett. **61**, 1029 (1988)
4. N. Read, S. Sachdev, Phys. Rev. Lett. **62**, 1694 (1989)
5. N. Read, S. Sachdev, Nucl. Phys. B **316**, 609 (1989)

6. T. Senthil, L. Balents, S. Sachdev, A. Vishwanath, M.P.A. Fisher, Phys. Rev. B **70**, 144407 (2004)
7. T. Senthil, A. Vishwanath, L. Balents, S. Sachdev, M.P.A. Fisher, Science **303**, 1490 (2004)
8. O.I. Motrunich, A. Vishwanath, Phys. Rev. B **70**, 075104 (2004)
9. A. Nahum, J.T. Chalker, P. Serna, M. Ortuño, A.M. Somoza, Deconfined criticality in a classical loop model (in preparation)
10. A. D'Adda, M. Lüscher, P. Di Vecchia, Nucl. Phys. B **146**, 63 (1978)
11. E. Witten, Nucl. Phys. B **149**, 285 (1979)
12. E. Fradkin, *Field Theories of Condensed Matter Physics*, 2nd edn. (Cambridge University Press, Cambridge, 2013)
13. I. Affleck, Nucl. Phys. B **257**, 397 (1985)
14. C. Candu, V. Mitev, T. Quella, H. Saleur, V. Schomerus, J. High Energy Phys. **02**, 015 (2010)
15. I. Affleck, Phys. Rev. Lett. **66**, 2429 (1991)
16. M. Kamal, G. Murthy, Phys. Rev. Lett. **71**, 1911 (1993)
17. A.W. Sandvik, Phys. Rev. Lett. **98**, 227202 (2007)
18. A.W. Sandvik, Phys. Rev. Lett. **104**, 177201 (2010)
19. F.J. Jiang, M. Nyfeler, S. Chandrasekharan, U.J. Wiese, J. Stat. Mech. **P02009**, 050405 (2008)
20. A. Banerjee, K. Damle, F. Alet, Phys. Rev. B **82**, 155139 (2010)
21. M. Matsumoto, C. Yasuda, S. Todo, H. Takayama, Phys. Rev. B **65**, 014407 (2001)
22. G. Murthy, S. Sachdev, Nucl. Phys. B **344**, 557 (1990)
23. M.A. Metlitski, M. Hermele, T. Senthil, M.P.A. Fisher, Phys. Rev. B **78**, 214418 (2008)
24. M.S. Block, R.G. Melko, R.K. Kaul, arXiv:1307.0519 (2013)
25. A.M. Polyakov, Phys. Lett. **59**, 82 (1975)
26. A.M. Polyakov, Nucl. Phys. B **120**, 429 (1977)
27. J.B. Kogut, Rev. Mod. Phys. **51**, 659 (1979)
28. M. Levin, T. Senthil, Phys. Rev. B **70**, 220403 (2004)
29. S. Powell, J.T. Chalker, Phys. Rev. Lett. **101**, 155702 (2008)
30. D. Charrier, F. Alet, P. Pujol, Phys. Rev. Lett. **101**, 167205 (2008)
31. S. Powell, J.T. Chalker, Phys. Rev. B **80**, 134413 (2009)
32. G. Chen, J. Gukelberger, S. Trebst, F. Alet, L. Balents, Phys. Rev. B **80**, 045112 (2009)
33. D. Charrier, F. Alet, Phys. Rev. B **82**, 014429 (2010)
34. R.K. Kaul, R.G. Melko, A.W. Sandvik, Annu. Rev. Condens. Matter Phys. **4**, 179 (2013)
35. K. Harada et al., Phys. Rev. B. **88**, 220408 (2013)

Chapter 4
The Statistics of Vortex Lines

4.1 Introduction

The loop models at $n = 1$ have the special property that the nodes are uncorrelated (in the absence of the extra interaction described in the previous chapter) so that the critical behaviour is purely geometrical. In this chapter we show that the corresponding field theories, the CP^0 and RP^0 models, have a broader applicability to the random geometry of topological vortex defects in random media.[1]

We will be concerned with problems of the following general kind. We are given a field $w(x)$, which is subject to either thermal or quenched disorder, and which can sustain vortex line defects. For example, if $w(x)$ is a random complex field, these are the lines $w(x) = 0$, around which the phase of $w(x)$ winds. A priori, vortices could either be suppressed, forming short closed loops, or they could proliferate and percolate through the system, and we may be able to access a continuous phase transition between the two regimes by tuning an appropriate parameter (examples are discussed below). We expect the geometrical structure of vortex lines to be universal whenever the typical vortex size diverges, and would like to know which field theories capture this universal behaviour.

Surprisingly, these field theories have not previously been provided, despite the fact that vortices play a crucial role in duality mappings for lattice models and field theories in three dimensions [1–7]. However there has been numerical work on the fractal structure of vortices in various contexts, motivated by cosmic strings [8–14], 'optical vortices' in disordered light fields [15, 16], the XY [17, 18] and Abelian Higgs models [19–23], turbulent superfluids [24] and, most usefully for us, a model for percolation with three colours [25–27].

A basic outstanding issue in relation to this numerical work is that of classification. It has been uncertain, for example, whether critical exponents for the geometrical vortex transition coincide with those for conventional percolation, whether the

---

[1] These universality classes also embrace 'deterministic walks in a random environment' such as the Lorentz lattice gas, and an important class of polymer models (Chap. 6).

A. Nahum, *Critical Phenomena in Loop Models*, Springer Theses,
DOI 10.1007/978-3-319-06407-9_4

universal behaviour is the same for oriented and unoriented vortex lines, and how the geometrical transition for vortices in the XY model relates to the conventional thermodynamic transition. The lack of a field theory description has also prevented a derivation of the most basic phenomenon, which is the appearance of Brownian statistics in the extended phase [8, 11, 15, 25–27], though this has previously been argued for by a heuristic analogy with polymers in the melt [11, 28].

The paradigmatic problem of geometry in disordered systems is of course percolation, which concerns domains in a random medium that are defined by some binary local property such as—in a continuum formulation—the sign of a random, short-range-correlated height field $h(x) \in \mathbb{R}$. In two dimensions, percolation may be viewed in terms of the 'line defects' which form the boundaries of domains, i.e. the zero lines $h(x) = 0$. They are of finite typical size when $h(x)$ is biased towards positive or negative values, but become scale-invariant random fractals at the critical point in between. Their universal properties are very robust, and show up in a surprising number of contexts ranging from the quantum Hall effect [29] to turbulence [30] to quantum chaos [31]. The fact that in two dimensions percolation can be viewed either in terms of domains (clusters) or in terms of loops leads to multiple field-theoretic descriptions: in addition to the $(q \to 1)$-state Potts model, which describes percolation in any number of dimensions, there is the description in terms of the CP^0 model with a θ term as well as the Coulomb gas description (reviewed in Chaps. 1 and 3).

In three dimensions, the geometry of line defects (which in general dimension D are associated with the homotopy group π_{D-2} of the order parameter manifold) can no longer be reduced to the geometry of domains (which are associated with π_0). As a result, there are more possibilities for universal behaviour, even in systems with short range correlations, than simply percolation.

The direct 3D analogue of the 2D level line problem above concerns the zero lines (vortices) of a complex field $w(x)$ with *short-range* correlations. Again, these vortices form a fluctuating soup of oriented loops, with the orientation of a loop defined by the sense in which the phase of $w(x)$ rotates around it. We may access a geometrical transition by varying the bias—the ratio of the average value of $w(x)$ to the width of its distribution [25–27]. We will consider this special case in detail. It will then be straightforward to generalise to other setups, including the (near-) critical XY model, and other order parameters such as nematics [9, 10, 13].

In the next section we give a heuristic explanation of the appearance of the CP^0 model, and relate this to the well-known duality between the XY model and the Abelian Higgs model [1–6]. We then give a more concrete lattice mapping for a specific model. (Reference [32] included an additional duality mapping in the continuum, but we omit this for brevity.) Finally we use the continuum description to answer some of the questions mentioned above.

4.2 Heuristic Considerations

Let us begin with a hand-waving argument for the form of the field theory for vortices in a random, short-range-correlated complex function $w(x)$.

It is useful to have in mind the phase structure that will emerge from our considerations, and which is seen numerically [14, 25–27]. When the probability distribution for $w(x)$ is $U(1)$ symmetric, or only weakly biased, vortices have a fractal dimension of two and a finite fraction of vortex density is in infinite vortex lines. This is the extended phase. A nonzero (translationally invariant) mean value $\langle w(x) \rangle$ suppresses vortices; when the mean is sufficiently large compared with the typical size of fluctuations, we enter the localized phase, and infinite vortices disappear. At the continuous transition between these phases, vortices are random fractals.

To see what field theory we should expect, let us view the loops as worldlines of quantum particles in $2+1$ dimensions. The Lagrangian governing the interactions of these particles must be chosen so that the partition function—which, expressed in the right basis, is a sum over worldline configurations—reproduces the statistical sum over vortex configurations.

Since vortices are oriented, we at first sight require a single charged boson z, but of course we will have to extend this either to an n-component vector $\mathbf{z}$, with the implied limit $n \to 1$, or to a supervector $\boldsymbol{\psi}$, in order to express geometrical correlation functions (Sect. 2.6). Global unitary or superunitary symmetry is then ensured by the fact that weight of a loop configuration does not depend on the colour indices which are thereby attached to loops, except by minus signs associated with fermionic colours.

It is tempting to assume that since $w(x)$ is short-range correlated, interactions between vortices will be short-range, and that we should therefore write down a field theory with only local interactions for $\mathbf{z}$. The natural candidate is then

$$\mathcal{L}_{\text{incorrect}} = |\nabla \mathbf{z}|^2 + \mu |\mathbf{z}|^2 + \lambda |\mathbf{z}|^4. \tag{4.1}$$

However, it is easy to see that this is not correct. For example, while the phase transition described by $\mathcal{L}_{\text{incorrect}}$ (at which $\mathbf{z}$ condenses) does represent a proliferation of $\mathbf{z}$ worldlines, this phase transition is thermodynamically nontrivial—it is in the XY universality class at $n = 1$—and so cannot represent the vortex transition we are discussing, where all local degrees of freedom are short-range correlated.

The error was of course in the assumption of short range interactions for $\mathbf{z}$. The weight for a vortex configuration $\mathcal{C}$ comes from integrating over all compatible configurations of $w(x)$, and this yields a long-range interaction between vortices.[2]

Note also that correlators of operators with nonzero $U(1)$ charge—such as $\langle z^\alpha(x) z^{*\alpha}(y) \rangle$—have no meaning in the vortex problem, since the number of vortex strands (worldlines) entering the vicinity of a point must equal the number leaving. This is a hint that $\mathbf{z}$ is coupled to a $U(1)$ gauge field, A, which both kills the unphysical correlators and simulates the long-range interaction between vortex strands[3]:

---

[2] However, a single vortex, once we 'integrate out' the configurations of the other vortices, has short range interactions with itself. In other words, the self-interactions of a vortex are screened by the other vortices (see Sect. 6.7).

[3] The gauge field also ensures that there are no massless degrees of freedom left when we set $n = 1$ in the replica theory, even when z is condensed—this is necessary, since there are no massless local degrees of freedom in the original problem.

$$\mathcal{L} = |(\nabla - iA)\mathbf{z}|^2 + \mu|\mathbf{z}|^2 + \lambda|\mathbf{z}|^4 + \mathcal{L}_A. \tag{4.2}$$

Again the partition function may be expressed as a sum over worldline (vortex) configurations: $\mathcal{L}_A$ must be chosen so that after integrating A out these configurations have the desired weights. (This is the basic idea of XY duality [1–6, 33], as discussed below.) Specifically, in the presence of the gauge field each vortex configuration $\mathcal{C}$ is weighted by the expectation value of a Wilson loop,

$$\mathcal{W}(\mathcal{C}) = \langle \exp i \int_{\mathcal{C}} A.\mathrm{d}x \rangle_A, \tag{4.3}$$

evaluated in the pure gauge theory. It is particularly easy to see what this weight should be in the limit of large bias. Suppose that $\langle w(x) \rangle$ is large and positive, and consider $\mathcal{W}(\mathcal{C})$ for a configuration with only a single large loop. Then $\arg w(x)$ will be close to zero for most x, but the vortex must bound a sheet where $\arg w(x) = \pi$. This sheet will cost an 'energy' scaling as the minimal area $\mathcal{A}$ enclosed by $\mathcal{C}$, so that $\log \mathcal{W}(\mathcal{C}) \sim -\mathcal{A}$.

This area law for the Wilson loop implies that A should be a compact gauge field [6, 34, 35]: we must include singular Dirac monopole configurations in the functional integral over A. (While we used large bias for this discussion, monopoles are present at any nonzero bias.) As discussed in Chap. 3, the compact gauge field confines $\mathbf{z}$ and $\bar{\mathbf{z}}$ into the gauge invariant composite Q. Finally, writing an effective action for Q leads us back to the CP^0 model.

This argument may be adapted to any order parameter with fundamental group $\pi_1 = \mathbb{Z}$. (In the above example, vortices are associated with $\pi_1(S^1) = \mathbb{Z}$, where S^1 is the space inhabited by $\arg w$.) This fundamental group guarantees the conservation of $U(1)$ charge in the field theory for $\mathbf{z}$, global $SU(n)$ (replica-like) symmetry relies only on the fact that vortices are oriented and topologically linear (see Sect. 4.4.1), and the argument for confinement requires only a bias favouring some point on the order parameter manifold.

We might however ask what happens in the absence of bias, or for order parameters whose fundamental group is different from $\mathbb{Z}$, or when correlations in $w(x)$ are not strictly short range. In Sect. 4.4.2 we show that different critical behaviour results when the fundamental group is $\mathbb{Z}_2$. Long range correlations may be addressed in a standard way using simple RG arguments (Sect. 4.4.3). Before considering what happens in the absence of bias, it is useful to rephrase the above argument in terms of XY duality, which gives a language in which it can be made semi-formal.

Consider the XY model for a planar spin $(\cos\theta, \sin\theta)$:

$$\mathcal{L}_{\mathrm{XY}} = \frac{K}{2}(\nabla\theta)^2 - H\cos\theta. \tag{4.4}$$

When the magnetic field H is nonzero, this Lagrangian yields one realisation of the problem discussed above, with a short-range, biased distribution for $w(x) \sim e^{i\theta(x)}$. As usual, the core fugacity for vortex defects is hidden in the ultraviolet regularisation.

When $H = 0$, a standard duality transformation relates the XY model to the Abelian Higgs model [1–6, 33]:

$$\mathcal{L}_{\text{dual}} = |(\nabla - iA)z|^2 + \kappa(\nabla \times A)^2 + \mu|z|^2 + \lambda|z|^4. \tag{4.5}$$

The worldlines of z represent vortices in $\theta(x)$, the potential for z encodes the length fugacity and short-range interactions of vortices, and the *non*-compact gauge field A mediates the long range interactions between vortices that (in the original language) arise from integrating out vorticity-preserving fluctuations in $\theta(x)$. The disordered phase of the XY model, in which vortices proliferate, is the Higgs phase of the gauge theory, where the field z has condensed and A is massive as a result of the Higgs mechanism. The ordered phase of the XY model is dual to the phase in which z is massive and A is massless (reflecting the presence of a massless Goldstone mode in the original description).

In order to discuss the bias-induced geometrical transition, we must extend the above duality in two ways. Firstly we must introduce replicated degrees of freedom ($z \to \mathbf{z}$) in order to capture geometrical information. Secondly, we must consider the theory (4.4) at nonzero H. In the dual language, the operators $e^{\pm i\theta}$ correspond to operators $\mathcal{M}$ and $\mathcal{M}^*$ that insert magnetic monopoles of positive or negative charge [6]. Therefore, the perturbation $H \cos\theta$ leads to a dual theory of the schematic form

$$\mathcal{L}_{\text{dual}} = |(\nabla - iA)\mathbf{z}|^2 + \kappa(\nabla \times A)^2 + \mu|\mathbf{z}|^2 + \lambda|\mathbf{z}|^4 + h\mathcal{M} + h\mathcal{M}^*, \tag{4.6}$$

with $h \propto H$ (see Ref. [32] for a more concrete discussion of duality at nonzero H). By expanding the Boltzmann weight $e^{-\int d^3x\, \mathcal{L}_{\text{dual}}}$ in h, we see that h is a fugacity for monopoles, so that A becomes a compact gauge field when $h \neq 0$. For scalar z, this perturbation would lead simply to a massive theory. However, for vector $\mathbf{z}$ we obtain a nontrivial theory when $h > 0$. This is equivalent to $\mathcal{L}_{\text{soft}}(Q)$ as far as universal properties are concerned.

Now consider the unbiased case, $H = 0$. Taking the stiffness K in Eq. 4.4 small, we realise a short-range correlated complex field without bias. In the dual picture, this corresponds to the Higgs phase in which $\mathbf{z}$ is condensed. Since gauge fluctuations are anyway massive in this phase, the fact that monopoles are absent at $H = 0$ does not lead to universal behaviour different to the rest of the extended phase. The duality does however suggest that $U(1)$ symmetry in the distribution of $w(x)$, together with short range correlations and appropriate spatial symmetries, is enough to *guarantee* that vortices are extended—we will return to this in Sect. 4.6.

We may also consider the behaviour of vortices in the vicinity of the XY critical point, which occurs at $H = 0$ and a critical stiffness $K = K_c$. At first sight we would expect this to lead to a new universality class for the geometry of vortex lines, governed by the critical 'NCCP^0' or 'NCCP$^{k|k}$' model (to adapt the terminology defined in Sect. 3.2.3). However it is likely (see Sect. 4.4.4) that in this case the thermodynamic XY transition is separate from the geometrical transition, and that the latter is again described by the (compact) CP^0 model. This is possible because the correlations in $w(x)$ do not need to be strictly short-range in order for the CP^0 universality class to apply.

Let us now turn to a lattice model for a random complex field [9–11, 25–27] in which the appearance of the dual degrees of freedom can be seen explicitly.

4.3 Tricolour Percolation

Figure 4.1 shows a tiling of space with cells of a certain shape. To generate a configuration of a discretised complex field $w(x)$, we randomly assign one of the three cube roots of unity, represented by the colours red, green and blue (for $w(x) = 1, e^{2\pi i/3}$, and $e^{4\pi i/3}$), to each cell x. The probabilities for the colours are p_R, p_G and p_B.

The cells are Wigner-Seitz cells for the bcc lattice, so this is site percolation with three colours on the bcc lattice. The edges of the cells form a second ('tetrakaidekahedral') lattice, which is four-coordinated, and this is where the vortices live. Each link of this lattice is surrounded by three cells, so we have a simple prescription for defining vortices: one is present on a link if the three adjacent cells take three different colours. The sense in which they appear around the link, or in other words the sense of rotation of the phase of $w(x)$, defines the orientation of the vortex. The structure of the lattice ensures that vortices are self- and mutually-avoiding.

This construction may seem like a brutal discretisation of the vortex problem, but we will see that it captures universal behaviour at the bias-induced transition. It has been studied numerically as a statistical mechanics problem in its own right [25–27], as an efficient means of generating configurations of self-avoiding polymers [25–27, 36, 37], and as a convenient setup for numerical investigations of cosmic string statistics [9–11].

Close to the symmetric point, $p_\mathrm{R} = p_\mathrm{G} = p_\mathrm{B} = 1/3$, vortices are in the extended phase [25–27]. Moving sufficiently far from this point, we enter the localized phase. Exponents are expected to be the same everywhere on the critical line. This line lies within the region in which all three colours percolate, so the vortex phase transition

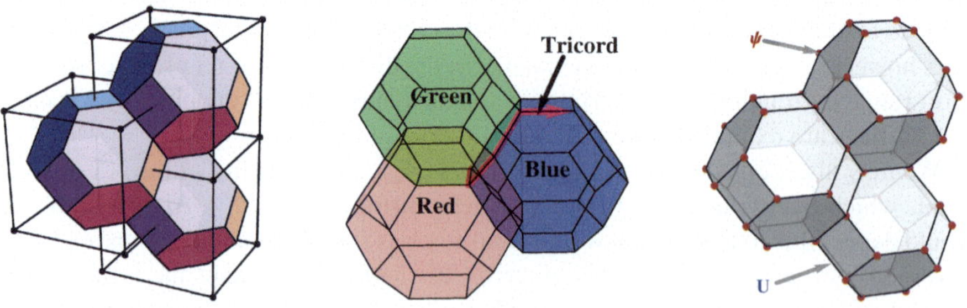

Fig. 4.1 *Left* Wigner-Seitz cells for the bcc lattice. *Centre* A section of vortex ('tricord'). *Right* Locations (sites and links) of the degrees of freedom ψ and U

does *not* coincide with the conventional percolation transitions for the colours. The numerical exponents of Ref. [25–27] are compatible with those for the $n = 1$ loop models (Chap. 1).

From our point of view, tricolour percolation is attractive because it allows a simple mapping to a lattice gauge theory. To see why this should be so, note that each configuration can be viewed as a random configuration of (branching) surfaces, which are just the domain walls between the colours. Ensembles of random surfaces appear naturally in the strong-coupling expansions of lattice gauge theories [38].

4.3.1 Symmetric Point

The lattice gauge theory takes its simplest form at the 'symmetric point' of the phase diagram, where all three colours have equal probability. Let i, j label sites of the tetrakaidekahedral lattice and let $\langle ij\rangle$ be the link from i to j. We introduce gauge fields U_{ij} on the links (unimodular complex numbers with $U_{ij} = U^*_{ji}$) and unit supervectors $\boldsymbol{\psi}_i$ on the sites,[4] defined as in Eq. 2.18. Then, letting F denote a face (where two cells meet) and l a link, the required partition function is

$$Z = \operatorname{Tr} \prod_F \left(1 + \prod_{l \in F} U_l + \prod_{l \in F} U_l^* \right) \prod_{\langle ij\rangle} \left(1 + U_{ij}^3 \boldsymbol{\psi}_i^\dagger \boldsymbol{\psi}_j + U_{ji}^3 \boldsymbol{\psi}_j^\dagger \boldsymbol{\psi}_i \right). \tag{4.7}$$

The links in the product $\prod_{l\in F}$ are oriented consistently around the face F. 'Tr' denotes the integral over $\boldsymbol{\psi}$ and U.

To see the relation to tricolour percolation, expand out the two products above, over faces F and over links $\langle ij\rangle$, and represent the terms graphically. The diagram for a given term is built up as follows. For each face F we must choose either 1, $\prod U$, or $\prod U^*$. If we choose '1' we add nothing to the diagram, while if we choose $\prod U$ or $\prod U^*$ we draw in the face F together with an orientation (equal or opposite to that of the links in the product respectively). We represent this orientation with a normal vector as in Fig. 4.2. For each link we must choose either 1, in which case we do not draw in the link, or $U_{ij}^3 \boldsymbol{\psi}_i^\dagger \boldsymbol{\psi}_j$, in which case we draw it in with an arrow from j to i, or $U_{ji}^3 \boldsymbol{\psi}_j^\dagger \boldsymbol{\psi}_i$, in which case we orient the arrow the other way.

After integrating over U, only those diagrams with an equal number of U_l and U_l^* on each link l survive. This leaves three possibilities, shown in Fig. 4.2. Either (I) neither the link nor any of the adjacent faces is included in the diagram; or (II) two faces are included, they are consistently oriented so as to form part of a sheet of oriented surface, and the link is not included; or (III) the link is included and so are all three faces, with orientations determined by the right hand rule from the orientation of the link. We are then left with sheets of oriented surface which close

[4] Since we are concerned only with loop fugacity $n = 1$, we temporarily use the SUSY formulation. But we can equivalently replace $\boldsymbol{\psi}$ with a unit vector $\mathbf{z}$, with $n \to 1$ implied.

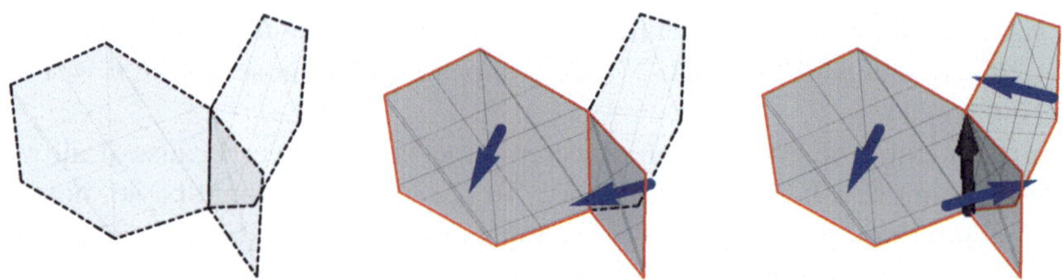

Fig. 4.2 Possibilities for the graphical expansion at a link. In the mapping to tricolour percolation, the *shaded surfaces* become domain walls between different colours

on themselves or meet along directed lines, which on a finite lattice form (self- and mutually-avoiding) loops.

The sheets of branching surface may be regarded as domain walls. Fixing the colour of one cell, we can colour the others by the rule that the colour changes cyclically, **R** → **G** → **B** → **R**, upon crossing a sheet in the same direction as its normal. The lines where three sheets branch are then the vortices. A convenient convention is to consider a finite bcc lattice and define the colour *outside* the lattice to be (say) red. Thus a cell on the boundary is red if its exterior faces (those on the boundary) are not covered by surface in the graphical expansion.

Let $\mathcal{T}$ denote a tricolour percolation configuration. At this point, having integrated over U but not $\boldsymbol{\psi}$, we have a correspondence analogous to (2.7):

$$Z = \sum_{\mathcal{T}} \operatorname{Tr} \prod_{\text{loops in } \mathcal{T}} (\boldsymbol{\psi}_1^\dagger \boldsymbol{\psi}_2) \cdots (\boldsymbol{\psi}_\ell^\dagger \boldsymbol{\psi}_1). \tag{4.8}$$

For each loop, the integral over the supervectors gives unity (i.e. the difference between the numbers of bosonic and fermionic colours—see Sect. 2.6). Z is therefore a sum over equally weighted tricolour percolation configurations, $Z = \sum_{\mathcal{T}} 1 = 3^{\text{no. cells}}$. Geometrical correlators may be constructed as in Chap. 2. For operators O_i that depend only on the $\boldsymbol{\psi}$s,

$$\langle O_1 \cdots O_N \rangle = \frac{1}{Z} \sum_{\mathcal{T}} \operatorname{Tr} O_1 \cdots O_N \prod_{\text{loops in } \mathcal{T}} (\boldsymbol{\psi}_1^\dagger \boldsymbol{\psi}_2) \cdots (\boldsymbol{\psi}_\ell^\dagger \boldsymbol{\psi}_1). \tag{4.9}$$

(The single-site integrals are given by a simple extension of Eq. 2.37,

$$\operatorname{Tr} \psi^{\alpha_1} \ldots \psi^{\alpha_q} \bar{\psi}^{\beta_1} \ldots \bar{\psi}^{\beta_q} = \frac{1}{q!} \left[\psi^{\alpha_q} \bar{\psi}^{\beta_1} \ldots \bar{\psi}^{\beta_q} \right]_{\mathrm{G}}, \tag{4.10}$$

where for the Gaussian average $\left[\psi^{\alpha} \bar{\psi}^{\beta} \right]_{\mathrm{G}} = \delta^{\alpha\beta}$.) For example the two-leg correlator may be written using $z^1 \bar{z}^2$ as in Chap. 2, though this operator is here defined at a site rather than a link.

The naive continuum limit of the lattice gauge theory (4.7) involves a superfield ψ coupled to a gauge field A. The compactness of the microscopic gauge degrees of freedom U_l translates in the continuum into the presence of Dirac monopoles in A [34, 35]. However, because ψ couples to U^3 rather than to U in (4.7), these monopoles are of three times the minimal charge allowed by the Dirac quantization condition. They are a result of threefold anisotropy in the distribution of $w(x)$ (see below). Unlike the monopoles of unit charge which appear in the next section, they do not play an important role, but for completeness we review how they arise.

We define the magnetic flux B_F passing through a given face (now taken to be oriented) by

$$\exp\left(\frac{iB_F}{3}\right) = \prod_{l\in F} U_l. \tag{4.11}$$

There is a 6π ambiguity in B_F, and thus also in its divergence $(\nabla.B)_x$, which is defined at each cell x by the sum of B_F over the outwardly oriented faces. This ambiguity can be resolved in two ways. One possibility is to restrict B_F to the domain $[-3\pi, 3\pi)$. This field then has nonzero divergence at the locations of monopoles: $(\nabla.B)_x = 2\pi\rho_x$, with monopole charges $\rho_x \in 3\mathbb{Z}$. Alternatively, we may resolve the ambiguity (call the new version B') in such a way that the divergence vanishes, $(\nabla.B')_x = 0$. B' differs from B by the inclusion of Dirac strings, which carry 6π flux to or away from the monopoles along strings of adjacent plaquettes on which B' lies outside the region $[-3\pi, 3\pi)$. These are lattice versions of the singular Dirac strings which carry flux away from monopoles in a continuum picture.

4.3.2 *Away from the Symmetric Point*

The basic objects in the graphical expansion of the lattice gauge theory are the domain walls between colours, rather than the colours themselves. At first sight this presents a problem if we wish to change the probabilities for the colours, but there is a simple way around this. To begin with, consider changing the probabilities for a single cell.

Recall that the colour *outside* the boundary is **R**. The colour of an interior cell x is therefore determined by the signed number of domain walls (modulo three) crossed by a path $\mathcal{P}$ from x to the boundary.

Introduce a variable ρ_x, which will be summed over $\rho_x = 0, \pm 1$; this will correspond to the charge of the monopole at x. Fix also a choice of path $\mathcal{P}$. Then, modify the Boltzmann weight for the lattice gauge theory in (4.7) by making the substitution $\prod U \to \exp\left(\pm\frac{2\pi i}{3}\rho_x\right)\prod U$ on the faces crossed by $\mathcal{P}$, where the sign depends on whether the plaquette's normal is parallel or antiparallel to $\mathcal{P}$.

The graphical expansion then goes through as before, except that for each term we can now read off the colour of x from the power of $\exp\left(\frac{2\pi i}{3}\rho_x\right)$ mod three. To control this colour, we augment the partition sum by a sum over ρ_x with the weight

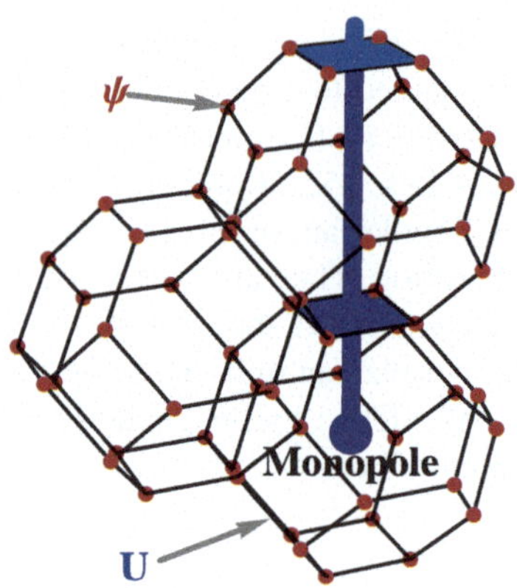

Fig. 4.3 Away from the symmetric point, monopoles of unit charge appear. In the presence of a monopole the Boltzmann weight for the gauge field is modified on the plaquettes crossed by the Dirac string

$$\sum_{\rho_x=0,\pm1} \left(p_{\mathrm{R}} + p_{\mathrm{G}} e^{2\pi i \rho_x/3} + p_{\mathrm{B}} e^{-2\pi i \rho_x/3}\right)(\cdots), \tag{4.12}$$

where the ellipsis stands for the other factors in the partition function. Performing the sum over ρ_x then associates the desired probability with each configuration.

For brevity, let us specialise to the line $p_{\mathrm{R}} \geq p_{\mathrm{G}} = p_{\mathrm{B}}$. The weight in the first brackets in (4.12) is then equal to $h^{|\rho_x|}$ with $h = (3/2)(p_{\mathrm{R}} - 1/3)$; h is a fugacity for monopoles, and vanishes at the symmetric point. To see the effect of nonzero h on the gauge theory, consider the Boltzmann weight in terms of gauge field configurations, rather than in terms of the graphical expansion. The Boltzmann weight for the plaquettes on $\mathcal{P}$ is now maximized not by a magnetic flux $B_F = 0$ (mod 6π), but by a flux $B_F = 2\pi\rho_x$ (mod 6π) oriented antiparallel to $\mathcal{P}$. Thus $\mathcal{P}$ is a Dirac string for a monopole of charge ρ_x located at x (Fig. 4.3).

The above argument extends immediately to a uniform (rather than local) change in the probabilities. Each cell acquires a monopole charge ρ_x, and we write

$$Z = \sum_{\{\rho_x\}} h^{N_\rho} \operatorname{Tr} \prod_F \left(1 + e^{i(B_F - D_F)/3} + \text{c.c.}\right) \prod_{\langle ij \rangle} \left(1 + U_{ij}^3 \boldsymbol{\psi}_i^\dagger \boldsymbol{\psi}_j + \text{c.c.}\right). \tag{4.13}$$

Here N_ρ is the total number of monopoles, $N_\rho = \sum_x |\rho_x|$. B_F is the magnetic flux through the face F, defined in (4.11), and D_F is the flux through F due to the Dirac strings $\mathcal{P}$ emanating from the monopoles. The geometry of the Dirac strings may be fixed by any suitable convention, subject to $(\nabla . D)_x = -2\pi\rho_x$.

Equation 4.13 is the desired lattice analogue of the field theory in Eq. 4.6. As a result of threefold anisotropy in $\arg w(x)$, we also have tripled monopoles, corresponding to a perturbation $\mathcal{M}^3 + \mathcal{M}^{*3}$ to the Lagrangian, even when $h = 0$. However these do not change the universal behaviour. A peculiarity of the above mapping is that monopoles appear in two distinct ways: as a result of the microscopic compactness of U and as a result of a modification to the Boltzmann weight for the gauge field (in a continuum derivation all monopoles appear in the same way [32]).

4.3.3 Wilson Loops

To connect up with the heuristic discussion in Sect. 4.2, we expand out only the second product (that over links) in the partition function (4.13). The diagrams then consist only of oriented links, to each of which is attached a factor $U^3 \psi^\dagger \psi$. Next, integrating over ψ kills all configurations except those in which the links form closed loops. Let $\mathcal{C}$ denote such a configuration of loops. Finally, the integral over U weights each 'worldline' configuration $\mathcal{C}$ by the corresponding Wilson loop:

$$Z \propto \sum_{\mathcal{C}} \mathcal{W}(\mathcal{C}) \operatorname{Tr} \prod_{\substack{\text{links}\langle ij\rangle \\ \text{in } \mathcal{C}}} (\psi_i^\dagger \psi_j) \qquad \mathcal{W}(\mathcal{C}) = \left\langle \prod_{\text{links } l \text{ in } \mathcal{C}} U_l^3 \right\rangle_U . \tag{4.14}$$

$\mathcal{W}(\mathcal{C})$ is evaluated using the part of the Boltzmann weight that depends only on the gauge field. This expectation value assigns a nontrivial entropic weight to each vortex configuration, given by the number of compatible colourings of the cells. (It vanishes for loop configurations which are not allowable vortex configurations, including those with intersections of loops.)

4.4 Consequences of the CP^0 Description

The CP^0 description implies that the universal behaviour discussed in Chap. 2 in the context of the $n = 1$ loop models applies also to (oriented) vortices.

It is possible to treat critical behaviour in a $6 - \epsilon$ expansion using the soft spin formulation of the field theory. Inserting the relevant constants in the formulas of Refs. [39–42] yields

$$\nu = \frac{1}{2} + \frac{5\epsilon}{76} + \frac{733\epsilon^2}{27436} + \cdots, \quad \eta = -\frac{\epsilon}{19} - \frac{166\epsilon^2}{6859} + \cdots . \tag{4.15}$$

(It is simplest to use the replica language. Writing $Q = q_i T_i$, where the q_i, $i = 1, \ldots, [n^2 - 1]$, are scalar fields and the T_i are $SU(n)$ generators, and dropping terms that are irrelevant in $6 - \epsilon$ dimensions, the field theory takes the form [40]

$$\mathcal{L}_{\text{soft}} = \frac{1}{2}(\nabla q)^2 + \frac{t}{2} q^2 + \frac{g}{4} d_{ijk} q_i q_j q_k . \tag{4.16}$$

The tensor d_{ijk} is defined by $\{T_i, T_j\} = \delta_{ij}/n + d_{ijk} T_k$. The coefficients in the epsilon expansion are given by contractions of the ds.)

In three dimensions, the expressions above give $\nu \simeq 0.94$ and $d_f \simeq 2.69$ ($\eta = -0.38$), to be compared with the values in Sect. 2.8.1 ($\nu = 0.9985(15)$, $d_f = 2.534(9)$ [43]). Since the $O(\epsilon^2)$ terms above are of similar size to the $O(\epsilon)$ terms at $\epsilon = 3$, some kind of resummation would be needed to obtain accurate results. The existence of a nontrivial fixed point in $6-\epsilon$ dimensions does however confirm that the upper critical dimension is six. (Recall that the cubic term vanishes at $n=2$, leading to an upper critical dimension of four; conceivably, we might have worried that something similar could happen at $n=1$, where Q itself vanishes.) The expressions (4.15) are compatible with $O(\epsilon)$ results for class C localization [44], as we would expect from the localisation mapping for the loop models.

The above also makes clear that, despite previous suggestions [9, 10, 25–27, 45], the critical exponents of vortices in tricolour percolation are distinct from those of conventional percolation, although small anomalous dimensions η in both problems lead to similar values for the fractal dimension of a critical vortex and of a critical percolation cluster; the latter has $d_f^{\text{perc}} = 2.5226(1)$ [46].

However, an appropriate perturbation will lead to a crossover to percolation as we now discuss.

4.4.1 Crossover to Percolation

In order to see the CP^0 exponents it is crucial that vortices are topologically one-dimensional. This issue arises naturally in lattice models (such as the XY model on a cubic lattice) where intersections of vortices are not forbidden. If intersections are resolved randomly in the same way that the nodes in the loop models were resolved, we will see CP^0 behaviour. But if the intersecting strands are fused together to form branching, net-like clusters rather than loops, we will see conventional percolation statistics for these clusters, as we now argue.

(The cancellation between bosonic and fermionic loops does not extend to clusters, so the following perturbation cannot be expressed in the supersymmetric language—we continue to use that of replicas.)

In the replica approach, the crossover to percolation is induced by a breaking of the global symmetry from $SU(n)$ down to the permutation group S_n.[5] This is the symmetry of the n-state Potts model which, in the limit $n \to 1$, is a well-known description of percolation [47].

Fused vortices must be assigned the same colour index, in order that the connectedness two-point function can be expressed via the colour-colour correlation function $\langle Q^{\alpha\alpha} Q^{\alpha\alpha} \rangle$ (Sect. 2.5). The perturbation induced by the fusing is then $\delta\mathcal{L} \sim -\sum_\alpha |z^\alpha|^4$, i.e. the operator which creates a meeting of four strands of the same colour. This perturbation can be written in terms of the 4-leg tensor (see Eq. 2.15 for the analog in the sigma model). However for our purposes it is sufficient to note that it yields a mass for the off-diagonal components of Q:

$$\delta\mathcal{L}_{\text{soft}} \sim \sum_{\alpha \neq \beta} |Q^{\alpha\beta}|^2. \tag{4.17}$$

[5] Together with $U(1)$ factors, which play no role since they act on massive fields.

The perturbation is expected, on the basis of its engineering dimension, to be strongly relevant. The minima of the resulting potential have Potts symmetry, so we expect percolation critical behaviour in the limit $n \to 1$. In the extended phase, the Goldstone modes acquire a mass.

4.4.2 Unoriented Vortices and Crossover to the RP^0 Model

So far we have discussed oriented vortices in random complex fields, which are associated with the fundamental group $\pi_1(S^1) = \mathbb{Z}$. It is natural to expect unoriented vortices, which appear when the relevant fundamental group is equal to $\mathbb{Z}_2$, to correspond to a similar field theory with real fields—the RP^{n-1} model—since the distinction between positively and negatively charged fields ($\mathbf{z}$ and $\bar{\mathbf{z}}$) relies on the orientation of the loops.

A quick way to confirm this is to view the unoriented vortex problem as a perturbation of the oriented one. Consider a nematic order parameter $\boldsymbol{v} = (v_1, v_2, v_3)$ with $\boldsymbol{v}^2 = 1$ and $\boldsymbol{v} \sim -\boldsymbol{v}$; the space of such vectors is RP^2, and $\pi_1(\mathrm{RP}^2) = \mathbb{Z}_2$. However the subspace $v_3 = 0$ is $\mathrm{RP}^1 = S^1$, so in the limit where fluctuations in v_3 are completely suppressed, we return to the oriented vortex problem. Turning fluctuations in v_3 back on allows the orientation of a vortex to fluctuate along its length. Specifically, the leading effect of reintroducing fluctuations in v_3 is to allow the configurations shown in Fig. 4.4 (right), where two vortices passing close to each other are rewired in a manner 'incompatible' with the orientations of the strands.

The corresponding perturbing operator creates two outgoing worldlines of the same colour and two incoming worldlines of the same colour: $\delta\mathcal{L} \sim -(\bar{z}^\alpha \bar{z}^\alpha)(z^\beta z^\beta)$. We see immediately that $SU(n)$ symmetry is broken to $SO(n)$. The perturbation has the same scaling dimension as that in Sect. 4.4.1, since the operators may be written in terms of the same irreducible tensor. In the soft spin theory, it yields a mass for the imaginary part of Q:

$$\delta\mathcal{L}_{\text{soft}} \sim \sum_{\alpha,\beta=1}^{n} (\mathrm{Im}\, Q^{\alpha\beta})^2. \tag{4.18}$$

Thus we expect a crossover to the RP^{n-1} model at $n = 1$, which is a similar theory for real Q. (This RP^{n-1} field should not be confused with the RP^2 field which hosts the vortices.) In Chap. 6 we will show that a similar crossover is important for 2D loop models.

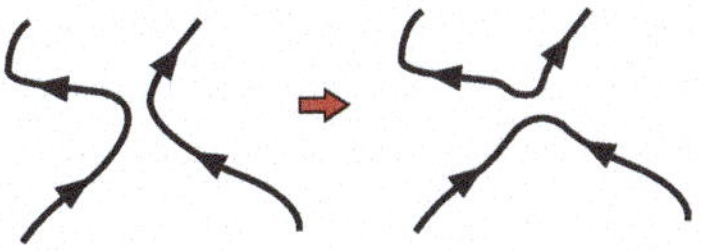

Fig. 4.4 'Rewiring' of vortices induced by perturbing the oriented vortex problem

4.4.3 Long Range Correlations

So far we have considered line defects only in short-range correlated random environments. How robust is this universal behaviour to long-range correlations? For weak correlations in $w(x)$, this may be addressed in a standard way using the extended Harris criterion [47, 48]. We view the correlations in $w(x)$ as due to quenched, correlated disorder in the parameters of the probability distribution for $w(x)$, e.g. for p_R, p_G and p_B in tricolour percolation, and average over this disorder to obtain an effective CP^0 Lagrangian with nonlocal couplings. The disorder-average does not require an additional use of the replica trick, since the partition function takes a trivial, disorder-independent value.

Consider correlations in $w(x)$ that decay as $(\text{distance})^{-A}$. The extended phase is extremely robust: Brownian behaviour is unaffected for any $A > 0$. (This seems to be compatible with the simulations of [12].) The critical theory on the other hand is stable so long as

$$A > \frac{2}{\nu} \simeq 2.003(3). \tag{4.19}$$

This is for generic correlations in $w(x)$. Cases involving oscillating correlations or Goldstone modes require separate comment.

Firstly, consider a random superposition of plane waves $\exp(ik.x)$ with fixed $|k|$, such as appears in Berry's random wave model [49] for eigenfunctions of the Laplacian in chaotic d-dimensional domains. We may fix $|k| = 1$ without loss of generality:

$$w(x) = \int_{k^2=1} \mathrm{d}^{d-1}k \; a(k) e^{ik.x}. \tag{4.20}$$

The complex numbers $a(k)$ are taken to be Gaussian and delta-correlated.

The 2D version of this problem with real $w(x)$ has been studied extensively in the context of quantum chaos, and the nodal lines $w(x) = 0$ are known to have the statistics of percolation cluster boundaries [50]. To reconcile this with the Harris criterion, it is necessary to take into account the oscillations in the correlator $\langle w(x)w(y)\rangle \sim |x-y|^{-1/2} \cos[|x-y| - \pi/4]$, as was pointed out in Ref. [31]. For complex $w(x)$ in 3D, similar considerations show that the collapse transition for the lines $w(x) = h$, induced by varying h, should be in the CP^0 universality class despite the slow decay of the correlator $\langle \bar{w}(x)w(y)\rangle \sim |x-y|^{-1} \sin|x-y|$.

Secondly, let $w(x) \sim e^{i\theta(x)}$ be the local order parameter for an XY model in its ordered phase. For weak long-range order in $w(x)$, vortices can remain in the extended phase, with a geometrical transition occurring as long range order gets stronger. At this transition, the phase θ exhibits slowly-decaying $1/(\text{distance})$ correlations due to Goldstone modes. However the relevant value of A for (4.19) is given by the correlator of the $O(2)$-symmetric operator $(\nabla\theta)^2$. This gives $A = 6$, so CP^0 behaviour is robust here too.

4.4.4 Vortices in the XY Model

Let us consider the near-critical XY model in more detail. Qualitatively, the XY transition may be viewed as a proliferation of vortex defects, as the conventional dual theory (4.5) makes clear. However, since this theory does not allow the expression of geometrical correlators, it does not shed light on whether the thermodynamic transition coincides with the geometrical one at which (appropriately defined) vortices begin to percolate.[6] At first sight, the replica-augmented dual theory,

$$\mathcal{L}_{\mathrm{NCCP}^{n-1}} = |(\nabla - iA)\,\mathbf{z}|^2 + \kappa(\nabla \times A)^2 + \mu|\mathbf{z}|^2 + \lambda|\mathbf{z}|^4 \quad (n \to 1) \tag{4.21}$$

or alternatively the $\mathrm{NCCP}^{k|k}$ model,

$$\mathcal{L}_{\mathrm{NCCP}^{k|k}} = |(\nabla - iA)\boldsymbol{\psi}|^2 + \kappa(\nabla \times A)^2 + \mu|\boldsymbol{\psi}|^2 + \lambda|\boldsymbol{\psi}|^4 \tag{4.22}$$

shows that the two transitions do coincide, occuring when $\mathbf{z}$ or $\boldsymbol{\psi}$ condenses. (There are no monopole operators in these Lagrangians since we are at zero magnetic field.) However this inference is not correct: in the presence of the non-compact gauge field, we must distinguish between the breaking or replica or supersymmetry, denoted $\langle Q \rangle \neq 0$, and the onset of the Higgs mechanism, denoted $\langle \mathbf{z} \rangle \neq 0$.

Instead, there is some numerical support [17, 18, 51] for the scenario in which the geometrical transition lies within the XY ordered phase. Since we can think of long range order as providing a nonzero bias, we would then expect the same critical behaviour as for the bias-induced transition. The above analysis shows that this is consistent with the Harris criterion.

In terms of field theory, the intermediate phase in which XY order and extended vortices coexist is a 'pair condensate' [54] with $\langle Q \rangle \neq 0$ but $\langle \mathbf{z} \rangle = 0$. In the vicinity of the vortex transition we expect to be able to write

$$\mathcal{L} = \kappa\,(\nabla \times A)^2 + \mathcal{L}_{\mathrm{soft}}[Q]. \tag{4.23}$$

The geometrical degrees of freedom are represented by Q, which vanishes in the replica limit. They decouple from the field A, which is the dual description of the Goldstone mode in the XY model, because of the irrelevance of the lowest

---

[6] Previous attempts have been made to relate the fractal structure of vortices in critical XY or Abelian Higgs models to local correlators in field theory using the duality between these two theories [20, 51, 52]. However, the authors did not make use of SUSY or replica, which are necessary in order to form nontrivial geometrical correlators. As a result the scaling relations put forward were not correct, as has previously been noted [21, 22, 53].

symmetry-allowed coupling $(\nabla \times A)^2 \operatorname{tr} Q^2$; this is equivalent to the condition from the Harris criterion.

Note that vortices in Abelian Higgs models, with Lagrangian $\mathcal{L}_{\text{SC}} = |(\nabla - ia)\Phi|^2 + \tilde{\kappa}(\nabla \times a)^2 + V(\Phi)$, can be treated in a similar manner to vortices in ungauged fields, leading to a dual replica/SUSY theory *without* a gauge field.

4.5 Aside: 2D Percolation

Above we used the fact that tricolour percolation was a problem of random surfaces to relate it to a lattice $U(1)$ gauge theory. This logic may also be applied to two-dimensional percolation. There, the 'random surfaces' are simply clusters of one colour: see Fig. 1.2, which may be taken to be a site percolation configuration on the faces of the honeycomb lattice. The loops are again oriented, if we take them to encircle the black clusters anticlockwise.

The required gauge theory is similar to Eq. 4.7, with the spins living on the sites of the honeycomb: we just need to alter the gauge field part of the Boltzmann weight to $\prod_F \left((1-p) + p\prod_{\langle ij\rangle \in F} U_{ij}\right)$. Note that we now have $\prod_{l\in F} U_l$ in the Boltzmann weight but not its complex conjugate—this is because there is only one kind of shaded face, unlike in the 3D model where shaded faces came with two orientations.

Coarse-graining this leads to the $\text{CP}^{k|k}$ model with a θ term (the SUSY analogue of Eq. 3.1), with $(\theta - \pi) \sim (p - \frac{1}{2})$. To see the appearance of the θ term, note that the product of Us around a face is the flux through that face, $\prod_{\langle ij\rangle\in F} U_{ij} \sim \exp\left[i\int_F \mathrm{d}^2x(\epsilon_{\mu\nu}\partial_\mu A_\nu)\right]$. A straightforward expansion in A gives

$$\prod_F \left[(1-p) + p\prod_{\langle ij\rangle\in F} U_{ij}\right] \sim \exp\int \mathrm{d}^2x \left(ip\,\epsilon_{\mu\nu}\partial_\mu A_\nu - \frac{p(1-p)}{2}(\epsilon_{\mu\nu}\partial_\mu A_\nu)^2 + \cdots\right).$$

The first term is the θ term (see Sect. 3.2.3). The fact that $\theta = \pi$ when $p = 1/2$ is protected by symmetry.

Setting $p = 1/2$, introducing a length fugacity x in front of the $U\psi^\dagger\psi$ terms in the partition function, and varying the numbers of bosons and fermions in ψ, we can access the full phase diagram of the honeycomb lattice loop model (Sect. 1.2).

We have arrived at a connection between loops on the honeycomb lattice and the CP^{n-1} model similar to that made by Affleck [55]—apart from the use of supersymmetry, the only (minor) difference in the above is the truncated form of the Boltzmann weight for the gauge field, which leads to a simpler graphical expansion. This is perhaps the most direct way to see the relationship between 2D loops without crossings and the CP^{n-1} or $\text{CP}^{n+k-1|k}$ model [55, 56]. (A lattice CP^{n-1} model with a different type of graphical expansion is described in Appendix A.)

4.6 Conclusion

Line-like topological defects are ubiquitous in three-dimensional systems, and continue to be of theoretical interest—for example, an exciting recent development is the realization that vortices in certain quantum systems can sustain topologically protected zero-energy states [57]. Here we have addressed the universal fractal geometry of vortices in disordered systems, a topic that has been studied numerically in diverse contexts but has lacked a field-theoretic description.

Let us summarise the possibilities for percolation-like transitions in 3D generally, where by 'percolation-like' we mean transitions in the geometry of domains or defects in short-range correlated random media.

For domains or clusters with no restriction on their topology, or the surfaces that form the boundaries of such domains [58], the relevant universality class is the familiar one of percolation and the continuum description is the $n \to 1$ state Potts model. For line defects that are strictly one-dimensional, i.e. do not branch, we have seen that the S_n symmetry of this Potts model is promoted to a continuous symmetry: the $SU(n)$ symmetry of the CP^{n-1} model for oriented line defects, and the $SO(n)$ symmetry of the RP^{n-1} model for unoriented line defects. The arguments of Sects. 4.2, 4.4.3, and 4.4.4 lead us to expect the CP^0 and RP^0 models to apply very generically to such line defects, so long as branching is not allowed, even if for example correlations are not strictly short range.

A natural question, which we will not answer fully here, is whether the three field theories mentioned above exhaust the possibilities for percolation-like transitions in 3D (according to the above definition). Can random fields with fundamental groups different from $\mathbb{Z}$ or $\mathbb{Z}_2$ yield new universality classes, or do we return to conventional percolation behaviour as soon as the lines can branch? A simple example to consider is an ensemble of oriented line defects associated with $\pi_1 = \mathbb{Z}_p$, in which p-fold branchings are allowed (where all the lines point into, or all point out of, the branch point). It is easy to see that if $p > 2$, coarse-graining generates the perturbation discussed in Sect. 4.4.1 (simply because we can make a four-leg vertex from two p-leg ones). Thus we expect to return to conventional percolation except in the case $p = 2$.

Incidentally, the patterns of symmetry breaking discussed here are relevant to other loop ensembles, not just line defects in 3D (see e.g. Sect. 5.3.7).

In a higher number D of dimensions, we expect the CP^0 and RP^0 models to describe oriented and unoriented line defects—for example, the zero lines of a random $(D-1)$-component field give an example of the oriented case. But we may also wonder whether the extended defects associated with other homotopy groups yield distinct universality classes. For example in four dimensions there are surface defects associated with π_1, as well as the line defects associated with π_2 and the domains associated with π_0.

It would be interesting to try to establish some of the properties of line defects in 3D rigorously.[7] Looking for models in which it could be rigorously proved that

[7] Some rigorous results for tricolour percolation have appeared since this thesis was written [59].

line defects percolate would be a natural starting point, because the duality mappings discussed in this chapter suggest that this property can be guaranteed, independently of microscopic specifics, by short range correlations in $w(x)$, together with $U(1)$ or $\mathbb{Z}_N$ symmetry in the distribution of $w(x)$ (e.g. the $\mathbb{Z}_3$ symmetry of tricolour percolation at the symmetric point).

In the next chapter, we will see that lattice gauge theory ideas similar to those discussed here lead to new kinds of universal behaviour in 2D loop models.

References

1. T. Banks, R. Myerson, J. Kogut, Nucl. Phys. B **129**, 493 (1977)
2. M. Stone, P.R. Thomas, Phys. Rev. Lett. **41**, 351 (1978)
3. M.E. Peskin, Ann. Phys. **113**, 122 (1978)
4. C. Dasgupta, B.I. Halperin, Phys. Rev. Lett. **47**, 1556 (1981)
5. M.P.A. Fisher, D.H. Lee, Phys. Rev. B **39**, 2756 (1989)
6. H. Kleinert, *Gauge Fields in Condensed Matter* (World Scientific, London, 1989)
7. M. Levin, T. Senthil, Phys. Rev. B **70**, 220403 (2004)
8. T. Vachaspati, A. Vilenkin, Phys. Rev. D **30**, 2036 (1984)
9. M. Hindmarsh, K. Strobl, Nucl. Phys. B **437**, 471 (1995)
10. K. Strobl, M. Hindmarsh, Phys. Rev. E **55**, 1120 (1997)
11. R.J. Scherrer, J.A. Frieman, Phys. Rev. D **33**, 3556 (1986)
12. R.J. Scherrer, A. Vilenkin, Phys. Rev. D **56**, 647 (1997)
13. T.W.B. Kibble, Phys. Lett. B **166**, 311 (1986)
14. T. Vachaspati, Phys. Rev. D **44**, 3723 (1991)
15. K. O'Holleran, M.R. Dennis, F. Flossmann, M.J. Padgett, Phys. Rev. Lett. **100**, 053902 (2008)
16. K. O'Holleran, M.R. Dennis, M.J. Padgett, Phys. Rev. Lett. **102**, 143902 (2009)
17. K. Kajantie, M. Laine, T. Neuhaus, A. Rajantie, K. Rummukainen, Phys. Lett. B **482**, 114 (2002)
18. E. Bittner, A. Krinner, W. Janke, Phys. Rev. B **72**, 094511 (2005)
19. S. Wenzel, E. Bittner, W. Janke, A.M.J. Schakel, Nucl. Phys. B **793**, 344 (2008)
20. J. Hove, S. Mo, A. Sudbø, Phys. Rev. Lett. **85**, 2368 (2000)
21. N. Prokof'ev, B. Svistunov, Phys. Rev. Lett **87**, 160601 (2001)
22. N. Prokof'ev, B. Svistunov, Phys. Rev. Lett **96**, 219701 (2006)
23. G. Vincent, N.D. Antunes, M. Hindmarsh, Phys. Rev. Lett. **80**, 2277 (1998)
24. D. Kivotides, C.F. Barenghi, D.C. Samuels, Phys. Rev. Lett. **87**, 155301 (2001)
25. R.M. Bradley, P.N. Strenski, J.-M. Debierre, Phys. Rev. A **45**, 8513 (1992)
26. R.M. Bradley, J.-M. Debierre, P.N. Strenski, Phys. Rev. Lett **68**, 2332 (1992)
27. R.M. Bradley, J.-M. Debierre, P.N. Strenski, J. Phys. A **25**, L541 (1992)
28. P.G. de Gennes, *Scaling Concepts in Polymer Physics* (Cornell University Press, Ithaca, 1979)
29. S.A. Trugman, Phys. Rev. B **27**, 7539 (1983)
30. D. Bernard, G. Boffetta, A. Celani, G. Falkovich, Nature Phys. **2**, 124 (2006)
31. E. Bogomolny, C. Schmit, J. Phys. A: Math. Theor. **40**, 14033 (2007)
32. A. Nahum, J.T. Chalker, Phys. Rev. E **85**, 031141 (2012)
33. S. Samuel, Nucl. Phys. B **154**, 62 (1979)
34. A.M. Polyakov, Phys. Lett. **59**, 82 (1975)
35. A.M. Polyakov, Nucl. Phys. B **120**, 429 (1977)
36. T. Prellberg, A.L. Owczarek, Phys. Rev. E **51**, 2142 (1995)
37. A. Bedini, A.L. Owczarek, T. Prellberg, Phys. Rev. E **86**, 011123 (2012)
38. J.B. Kogut, Rev. Mod. Phys. **51**, 659 (1979)
39. D.J. Amit, J. Phys. A **9**, 1441 (1976)

40. A.J. McKane, D.J. Wallace, R.K.P. Zia, Phys. Lett B **65**, 171 (1976)
41. O.F. de Alcantara Bonfim, J.E. Kirkham, A.J. McKane, J. Phys. A: Math. Gen **13**, L247 (1980)
42. O.F. de Alcantara Bonfim, J.E. Kirkham, A.J. McKane, J. Phys. A: Math. Gen **14**, 2391 (1981)
43. M. Ortuño, A.M. Somoza, J.T. Chalker, Phys. Rev. Lett. **102**, 070603 (2009)
44. T. Senthil, unpublished
45. A.M.J. Schakel, J. Low Temp. Phys. **129**, 323 (2002)
46. Y. Deng, H.W.J. Blöte, Phys. Rev. E **72**, 016126 (2005)
47. J. Cardy, *Scaling and Renormalization in Statistical Physics* (Cambridge University Press, Cambridge, 1996)
48. A. Weinrib, B.I. Halperin, Phys. Rev. B. **27**, 413 (1983)
49. M.V. Berry, J. Phys. A: Math. Gen. **10**, 2083 (1977)
50. E. Bogomolny, R. Dubertrand, C. Schmit, J. Phys. A **40**, 381 (2007)
51. M. Camarda, F. Siringo, R. Pucci, Phys. Rev. B **74**, 104507 (2006)
52. J. Hove, A. Sudbø, A. Phys. Rev. Lett. **96**, 219702 (2006)
53. W. Janke, A.M.J. Schakel, cond-mat/0508734 (2005)
54. O.I. Motrunich, A. Vishwanath, arXiv:0805.1494 (2008)
55. I. Affleck, Phys. Rev. Lett. **66**, 2429 (1991)
56. N. Read, H. Saleur, Nucl. Phys. B **613**, 409 (2001)
57. J.C.Y. Teo, C.L. Kane, Phys. Rev. B **82**, 115120 (2010)
58. R.M. Bradley, P.N. Strenski, J.-M. Debierre, Phys. Rev. B. **44**, 76 (1991)
59. S. Sheffield, A. Yadin, Electron. J. Probab. **19**, 1 (2014)

Chapter 5
Loop Models with Crossings in 2D

5.1 Introduction

It is tempting to think that two-dimensional critical phenomena are completely classified and understood, thanks to conformal field theory and other exact techniques, but this is far from true. One class of problems which remains mysterious is that containing classical loop (and polymer) models, together with models for non-interacting fermions subject to disorder. As discussed in Chap. 1, these systems are tied together by field theory descriptions with continuous replica-like symmetries or alternatively global supersymmetries [1–13].

Loop models are in many ways simpler than disordered fermions, but even they are not yet fully understood. The best-studied examples are those in which the loops are forbidden from crossing, such as the honeycomb lattice model or the completely-packed model on the 2D L lattice (Sects. 1.2, 2.2 and 3.2.1). For these a great deal is known from conformal field theory, height model mappings, exact solutions, Schramm-Loewner Evolution, and numerical simulations [14, 15]. But when we move away from these models the analytical techniques often cease to apply, and we may encounter new types of critical phenomena requiring new theoretical tools.

In this chapter we address some of the simplest 2D loop models with crossings. These reveal new universality classes of, and new mechanisms for, classical critical behaviour. They also provide natural models for polymers and for deterministic motion in a random environment [16–18] which have been intensely studied but whose phase diagrams and continuum descriptions have in general not been understood. Finally, they shed light on phenomena that are important more generally for criticality in replica or supersymmetric sigma models—in particular, the role of gauge symmetries and topological point defects. The latter have recently been shown also to be important for 2D Anderson metal-insulator transitions [19, 20]. We will return to the analogy with localisation at the end of this introduction.

A key result of previous work on loops with crossings is the existence of an unusual critical phase which is absent for non-crossing loops [10, 11, 16, 18, 21–23]. It was argued by Jacobsen, Read and Saleur [10, 11] that this corresponds

A. Nahum, *Critical Phenomena in Loop Models*, Springer Theses,
DOI 10.1007/978-3-319-06407-9_5

Fig. 5.1 A configuration of the completely packed loop model with crossings (CPLC) on a 10×10 lattice with periodic boundary conditions. Each loop has been given a different colour—these colours are merely a guide to the eye and not part of the configuration

to the *Goldstone* phase of the $O(n)$ sigma model, where as usual n is the fugacity for loops. The phase exists for $n < 2$. Characteristic features of it had previously been found in computational studies of polymers and deterministic walks in a random environment [16, 18], as well as in an integrable loop model [21–23]. It appears quite generically when non-crossing loop models in the dense regime are perturbed by the addition of crossings, which corresponds to a breaking of symmetry (Sect. 1.3.2).

Here we examine a more general class of loop models with crossings. These show continuous phase transitions in new universality classes, separating the Goldstone phase from non-critical phases with short loops. We construct field theories for them, and pin down the universal behaviour (both in the Goldstone phase and at the new critical points) using analytic calculations and extensive Monte Carlo simulations.[1]

The models we study are a 'completely-packed loop model with crossings' (CPLC) on the square lattice—Fig. 5.1 shows a configuration—and an 'incompletely packed loop model with crossings' (IPLC) in which loops are related to cluster boundaries. The parameter space for the CPLC contains various previously-studied models as special cases, including the standard completely-packed loop model without crossings and models with crossings encountered in various contexts. The CPLC and IPLC are expected to show the same universal behaviour. Our numerics are restricted to the CPLC, but the IPLC provides a simpler context in which to describe the main theoretical ideas.

It is easy to argue that the phase transitions in the CPLC and IPLC cannot be described by the $O(n)$ model. Instead, a general description of the models requires

[1] The Monte Carlo results reported here were obtained by Pablo Serna, Andres Somoza, and Miguel Ortuño. The approach to simulations and analysis of numerical results were developed jointly.

us to couple the $O(n)$ spin $\mathbf{S}$ to a $\mathbb{Z}_2$ gauge field, or equivalently to identify $\mathbf{S}$ with $-\mathbf{S}$, yielding a nematic order parameter. This leads to a sigma model on real projective space, RP^{n-1}, in which $\mathbb{Z}_2$ point vortices play an important role. Vortices are suppressed in the Goldstone phase—meaning that the $O(n)$ model is a viable description there—but proliferate at the phase transition into the short loop phase. (This picture is appropriate for the regime $0 < n < 2$.)

For the critical points we are restricted to numerics and approximate RG treatments (since the usual exact techniques rely on the absence of crossings) [20]. However the Goldstone phase can be understood in full analytically, since it is characterised by marginal flow to a weak-coupling fixed point [10, 11]. This leads to logarithms—e.g. correlation functions decaying with a universal power of the logarithm of distance—so very large system sizes are required in order to confirm our analytical predictions numerically (comparable to the largest simulated in any statistical mechanics problem). These are possible at fugacity $n = 1$ thanks to special features of the problem there, and our simulations are restricted to this value.

Another feature of the CPLC at $n = 1$ is that while each configuration is a soup of many loops, the model permits a mapping to a model for a *single* loop with local interactions. At a certain point in parameter space, this is the 'interacting self-avoiding trail' (ISAT) model for a polymer at its collapse, or Θ, point [16, 24] (see next chapter).

We have already mentioned the similarities between loop models and localisation [25–31]. Supersymmetry and replica-like limits, crucial in the latter for averaging over disorder, appear in the former as tools allowing the expression of geometrical correlation functions. Both types of problem exhibit critical points of central charge zero, described by logarithmic CFTs [32–35]. Loop models are a good place to study such critical points since they are more tractable, both analytically and computationally, than disordered fermion problems.

The relationship is closest for completely packed loops on the L lattice at fugacity $n = 1$, which are connected by an exact mapping [25, 26, 30, 36] to a network model for the spin quantum Hall transition [7, 37–40]. However the analogy is more general. Recent work by Fu and Kane [20] demonstrates that the metal-insulator transition in the symplectic symmetry class is driven by proliferation of $\mathbb{Z}_2$ vortices: this transition is thus in remarkably close analogy with those in the loop models discussed here, though the appropriate sigma model is different. In the localisation language, the Goldstone phase corresponds to a metallic phase, and the two short loop phases—which are distinguished from each other by the presence or absence of a loop encircling the boundary—to topological and trivial insulating phases.

The CPLC at $n = 1$ can in fact be obtained as a 'classical' limit of a network model in which a Kramers doublet propagates on every edge. The above similarities show that this classical limit captures a surprising number of the qualitative features of the phase diagram for the symplectic class (see also Sect. 5.7).

In both the loop model and the localisation problem the $\mathbb{Z}_2$ vortex fugacity, including its sign, plays an important role. Reference [20] introduced an approximate RG treatment of this fugacity, and in Sect. 5.5.3 we apply this to the loop models. Vortices—this time $\mathbb{Z}$ vortices—have also been shown to be responsible for Anderson localisation in the chiral symmetry classes, and a detailed treatment has been given by König et al. in Ref. [19].

We now introduce the models we study and their phase diagrams. In Sect. 5.3 we map them to lattice gauge theories and RP^{n-1} sigma models, paying attention to the role of topological defects. We also discuss the different (CP^{n-1}) sigma model which applies on the boundaries of the phase diagram for the CPLC, and make explicit the connection between topological features in the two sigma models. Section 5.4 treats the Goldstone phase via Monte Carlo and RG, and Sect. 5.5 treats the critical points. Our numerical methods are described in more detail in Sect. 5.6. Finally, Sect. 5.7 discusses directions for future work (see also Chap. 7).

5.2 Definitions of Models

5.2.1 Completely-Packed Loops with Crossings

The completely-packed loop model with crossings (CPLC) is defined in a similar way to the completely-packed loop models of Chap. 2 , except that the links of the lattice are no longer oriented and correspondingly there are three allowed pairings at each node rather than two. These are shown in Fig. 5.2. Figure 5.1 is an example of a configuration on a small lattice. Overcrossings are not distinguished from under-crossings: the configuration at a node is defined solely by the way the links are paired up.

The pairings are assigned the weights shown in Fig. 5.2, with that of a crossing being p. The factors q and $1-q$ are staggered (swapped) on the two sublattices of the square lattice, so that the states of the system for extreme values of the parameters are as shown in Fig. 5.3. (The phase diagram is a triangle, since the value of q is unimportant when $p = 1$: hence the rescaled coordinate $\widetilde{q}$ in this figure.)

Letting N_p, N_q and N_{1-q} denote the numbers of nodes where the pairings with weight p, $(1-p)q$ or $(1-p)(1-q)$ are chosen, the product of node weights in a configuration $\mathcal{C}$ is

$$W_{\mathcal{C}} = p^{N_p}\,[(1-p)q]^{N_q}\,[(1-p)(1-q)]^{N_{1-q}}, \qquad (5.1)$$

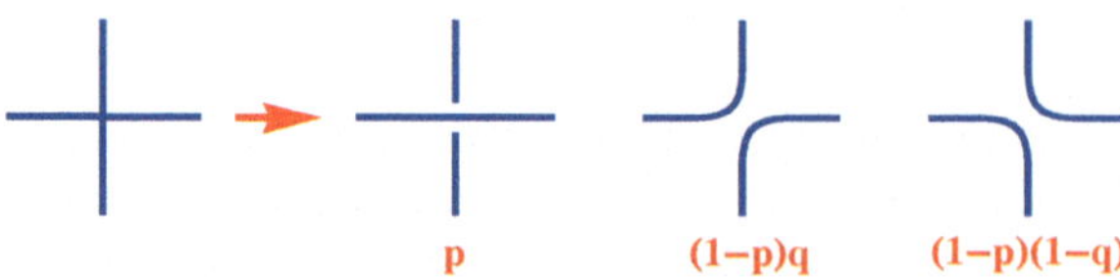

Fig. 5.2 The three configurations of a node and associated Boltzmann weights. (In the *leftmost* configuration, the *upper* and *lower* links lie on the same loop.) The weights q and $1-q$ are exchanged on the two sublattices of the square lattice

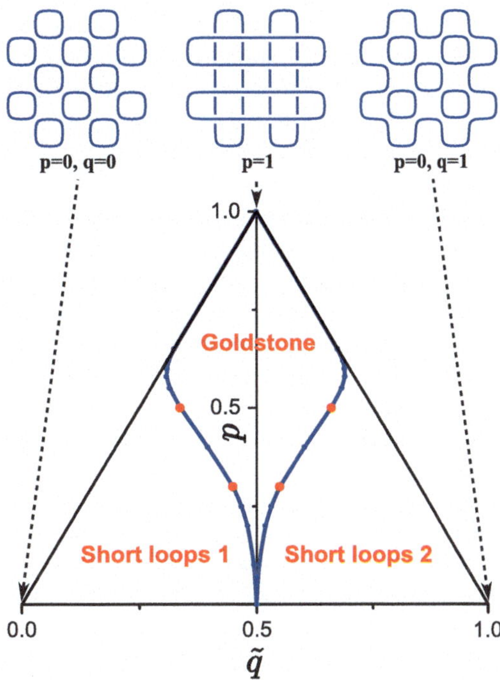

Fig. 5.3 Phase diagram obtained numerically for the CPLC at $n = 1$. The *horizontal* axis is labelled by $\tilde{q}$, defined by $(\tilde{q} - 1/2) = (q - 1/2)(1 - p)$. The *larger* (*red*) *dots* on the critical line indicate the values of p at which we have analysed the critical behaviour in detail. Also shown are the configurations obtaining on a small finite lattice at $p = 1$, at $p = 0$, $q = 0$, and at $p = 0$, $q = 1$. (The point $p = 0$, $q = 1/2$ is the percolation critical point)

and the partition function is defined in the usual way by

$$Z = \sum_{C} n^{\text{no. loops}} W_C. \tag{5.2}$$

The parameter space of this model includes various previously-investigated models. To begin with, note that whenever one of the node weights vanishes, the links of the lattice can be assigned fixed orientations (with two incoming and two outgoing links at each node) such that the allowed pairings are always between an incoming and an outgoing link. This orients all the loops. When $p = 0$, the required orientations are those of the L lattice (Fig. 2.1). Thus at $p = 0$, $q = 1/2$ we have the critical behaviour of the dense phase (note that in Fig. 2.3 the parameter q was instead called p). When either $q = 0$ or $q = 1$, we can orient the links as in Fig. 2.2, and we have the Manhattan lattice loop model studied in Refs. [26, 41].

The model on the line $q = 1/2$ was related to the Goldstone phase of the $O(n)$ sigma model in Ref. [11], and points on this line have been studied in various contexts. For a given value of n, the point $q = 1/2$, $p = (2-n)/(10-n)$ is known as the Brauer loop model [21–23] and is integrable; this model was related to a supersymmetric spin chain in Ref. [21]. Unfortunately, integrability gives only limited information. When the parameters are such that all configurations are given equal weight—i.e. when $n = 1$, $q = 1/2$ and $p = 1/3$—a simple mapping relates the CPLC to a standard model for polymers at their Θ (collapse) point [16, 17, 24], which we will discuss in the next chapter. Loop models with crossings at $n = 1$ have also appeared in the study of Lorentz lattice gases, i.e. deterministic motion in a random environment [18, 42]. Finally, Ref. [43] discusses a model similar to the CPLC in which q and $1-q$ are not staggered, and uses it at $n = 2$ to analyse the phase diagrams of vertex models.

A trivial but important fact about the CPLC is that (as in the models of Chap. 2) the nodes become completely independent of each other when $n = 1$. The weights p, $(1 - p)q$ and $(1 - p)(1 - q)$ are then the probabilities of the various node configurations, and the partition function Z is equal to unity—from which it follows, by the finite size scaling of the free energy, that any critical points must have central charge $c = 0$. The model with $n = 1$ is thus analogous to percolation, which can also be formulated in terms of uncorrelated random variables. In the absence of crossings the $n = 1$ model is in fact equivalent to bond percolation on a dual lattice, with loops surrounding cluster boundaries; however when crossings are allowed the universal behaviour is no longer that of percolation.

Our simulations will be restricted to the case $n = 1$, which is the most interesting and the best suited to Monte Carlo, but most analytic results will apply to $0 \leq n < 2$. (We will discuss the case $n = 2$, which shows more conventional critical behaviour, elsewhere.) The phase diagram obtained numerically at $n = 1$ is shown in Fig. 5.3. We expect it to be qualitatively similar for $0 < n < 2$, with the Goldstone phase swallowing up more and more of the parameter space as $n \to 0$. At the cost of repeating some of what we have already said, we now summarize its main features.

Short loop phases. The configurations at $p = 0, q = 0$ and at $p = 0, q = 1$ provide caricatures of the two 'short loop' phases. For given boundary conditions, these are distinguished from each other by the presence or absence of a long loop running along the boundary, as shown in Fig. 5.3. In a sigma model description the short loop phases are massive (disordered) phases. In the analogy with Anderson localization they correspond to insulating phases, and the boundary loops correspond to the edge states present in a topological insulator.

Goldstone phase. In the Goldstone phase, so-called because the continuum description is a sigma model which flows to weak coupling in the infra-red [11], the loops are 'almost' Brownian. However, interactions between Goldstone modes in the sigma model are only marginally irrelevant, leading to universal logarithmic forms for correlators and other observables which we calculate in Sect. 5.4. In the Anderson localization analogy, this would be a metallic phase.

Critical lines. The lines separating the Goldstone phase from the short-loop phases show a new universality class of critical behaviour. This is associated with the order-disorder transition of the RP^{n-1} sigma model, which exists only in the replica limit $n < 2$ and is driven by proliferation of $\mathbb{Z}_2$ vortex defects associated with $\pi_1(\mathrm{RP}^{n-1})$ (Sects. 5.3 and 5.5.3). Numerically, the critical loops have $d_f = 1.909(1)$ at $n = 1$, i.e. they are slightly less compact than Brownian paths, and the transition has a large correlation length exponent $\nu = 2.745(19)$ (Sect. 5.5).

Critical point at $p = 0, q = 1/2$. As already mentioned, the critical point of the loop model without crossings corresponds to the dense phase of non-crossing loops, or to SLE_κ with $\kappa > 4$ (Sect. 1.4.3). At $n = 1$ this critical point maps to critical percolation: the loops have the statistics of percolation cluster boundaries, with a fractal dimension $d_f^{\mathrm{perc}} = 7/4$, and the correlation length exponent of the transition is $\nu^{\mathrm{perc}} = 4/3$. These critical exponents also yield exponents in an Anderson transition (the spin quantum Hall transition) via an exact mapping [25, 26, 36, 44].

Phase diagram boundaries. On the boundaries of the phase diagram Fig. 5.3, we are really dealing with models of oriented loops (on either the L or the Manhattan lattice, depending on which of the node weights vanishes) as noted above. The fact that the loops automatically come with an orientation means that the continuum descriptions have a higher symmetry [9, 45]—the appropriate field theory is the CP^{n-1} model rather than the RP^{n-1} model that applies in the interior of the phase diagram. The CP^{n-1} description implies that the lines $q = 0$ and $q = 1$ (Manhattan) are always in the short loop phase, but with a typical loop length that diverges exponentially as $p \to 1$ (Sect. 5.3.6). This is in agreement with previous expectations [26, 41], but is not obvious from the numerical phase diagram since the critical lines closely approach the lines $q = 0, 1$ for p close to one.

Relation to polymers. Configurations in the CPLC are soups of many loops. However when $n = 1$ the CPLC has a simple relation with the self-avoiding trail model for a single polymer [16, 17]. The polymer corresponds to a single marked loop in the soup of loops; so long as $n = 1$, 'integrating out' the configurations of the other loops leads to a local Boltzmann weight for the marked one.[2] Adding the interactions that are natural in the polymer language takes us out of the parameter space of Fig. 5.3, but the sigma model description can be extended to cover this case by including appropriate symmetry-breaking terms (Chap. 6).

5.2.2 *Incompletely-Packed Loops with Crossings*

For pedagogical reasons, it will be useful to introduce and discuss a second loop model before returning to the CPLC. Loops in the new model will no longer be completely packed, but nevertheless the universal properties will be the same. We refer to this model as the incompletely-packed loop model with crossings, or IPLC.

To generate a configuration in the IPLC, we first colour the plaquettes of the square lattice black or white, giving a site percolation configuration on the square lattice formed by the plaquettes. The loops in the IPLC are then cluster boundaries, as shown in Fig. 5.4. However the loop configuration is not uniquely determined by the cluster configuration: for each 'doubly visited' node, where two cluster boundaries meet, we must choose how to connect them up. Allowing crossings, the three possible pairings are again those of Fig. 5.2 (but unlike in the CPLC we do not assign different weights to the different pairings).

---

[2] This locality is easy to see: since the nodes are independent, the probability that a loop $\mathcal{L}$ appears in a randomly chosen configuration is equal to a product of node factors for nodes $\mathcal{L}$ visits.

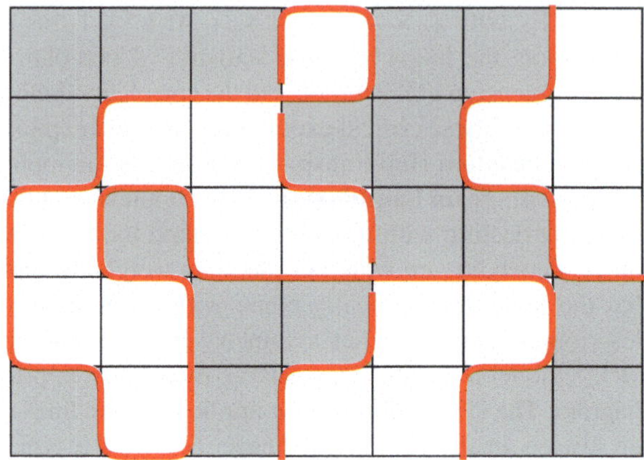

Fig. 5.4 Part of a configuration in the IPLC. In this model the loops (*thick red lines*) are cluster boundaries. Universal behaviour in the IPLC is expected to coincide with that in the CPLC

The simplest choice for the Boltzmann weight is to give each percolation configuration the standard percolation probability $\mathrm{q}^{\mathrm{B}}(1-\mathrm{q})^{\mathrm{W}}$, where B and W are the numbers of black and white faces. A given percolation configuration corresponds to 3^N loop configurations, where N is the number of doubly-visited nodes. Assigning them equal probability, the partition function for the IPLC is

$$Z = \sum_{\text{configs}} a^N \mathrm{q}^{\mathrm{B}}(1-\mathrm{q})^{\mathrm{W}}, \tag{5.3}$$

with $a = 1/3$. The parameter q plays similar role to the parameter q in the CPLC.

The above partition function corresponds to loop fugacity $n = 1$. We will wish to generalize it to arbitrary n. We may also vary a, and introduce a fugacity x for the total length of loops:

$$Z = \sum_{C} a^N \mathrm{q}^{\mathrm{B}}(1-\mathrm{q})^{\mathrm{W}} x^{\text{length}} n^{\text{no. loops}}. \tag{5.4}$$

The precise values of a and x will not be important in what follows. When $n = 1$, a length fugacity x distinct from one corresponds to an Ising interaction of strength J between colours of adjacent squares, with $x = e^{-2J}$, and varying a introduces a four-square interaction. These interactions have no effect on the universal behaviour so long as they are weak.

Note that if we take our lattice to have the topology of the disk, regarding the region outside the boundary as white for the purposes of drawing cluster boundaries, the IPLC shares with the CPLC the feature of having an edge loop at $\mathrm{q} = 1$ but not at $\mathrm{q} = 0$. There are clearly stable short-loop phases at small q and at q close to one, and field theory arguments lead us to expect a stable Goldstone phase near $\mathrm{q} = 1/2$.

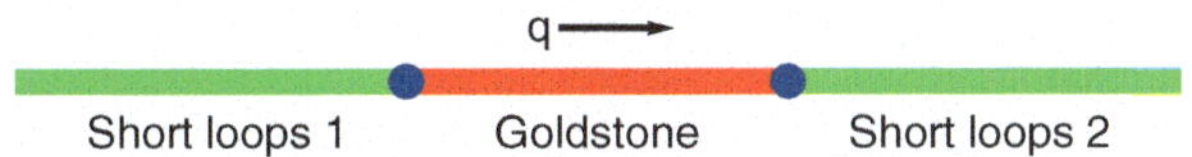

Fig. 5.5 Schematic phase diagram for the incompletely-packed loop model with crossings as a function of q (for fixed $n \sim 1, a \sim 1/3, x \sim 1$)

The conjectured phase diagram, shown in Fig. 5.5, is similar to a slice through the phase diagram of the CPLC at some nonzero value of p. (If we had forbidden crossings, we would have obtained a phase diagram similar to the line $p = 0$ in the CPLC.)

5.3 Lattice Field Theories

5.3.1 Lattice Field Theory for IPLC

The IPLC permits a simple mapping to a lattice model with a $\mathbb{Z}_2$ gauge field coupled to $O(n)$ spins. This is very similar to the mappings in Chap. 4 , but we repeat the steps so as to make this discussion self-contained.

The $O(n)$ spins $\mathbf{S}_i = (S_i^1, \ldots, S_i^n)$, $\mathbf{S}_i^2 = n$, live on the sites of the square lattice, and the gauge fields $\sigma_{ij} = \pm 1$ on the links. The partition function is:

$$Z = \mathrm{Tr} \prod_F \Big((1-\mathrm{q}) + \mathrm{q} \prod_{\langle ij \rangle \in F} \sigma_{ij} \Big) \prod_{\langle ij \rangle} \left(1 + x \sigma_{ij} \mathbf{S}_\mathbf{i} . \mathbf{S}_j \right). \tag{5.5}$$

F denotes a face (square) of the lattice. The $\mathbb{Z}_2$ gauge symmetry is

$$\mathbf{S}_i \to \chi_i \mathbf{S}_i, \quad \sigma_{ij} \to \chi_i \chi_j \sigma_{ij} \qquad (\text{for } \chi_i = \pm 1). \tag{5.6}$$

Above we have written the Boltzmann weight for the gauge field in a form suitable for the graphical expansion. Later we will rewrite it in a more conventional form.

We expand out the product over faces F in (5.5), colouring a face black or white depending on whether the q or the $1 - \mathrm{q}$ term is chosen. This generates percolation configurations $\mathcal{P}$,

$$Z = \sum_{\mathcal{P}} \mathrm{q}^{\mathrm{B}} (1-\mathrm{q})^{\mathrm{W}} \, \mathrm{Tr} \Big(\prod_{l \in \partial \mathcal{P}} \sigma_l \Big) \prod_{\langle ij \rangle} \left(1 + x \sigma_{ij} \mathbf{S}_i . \mathbf{S}_j \right), \tag{5.7}$$

where $\partial \mathcal{P}$ denotes the set of links on cluster boundaries. Next, expanding out the product over links and summing over σ_{ij}, the only surviving term is that in which the factors of $\mathbf{S}_i.\mathbf{S}_j$ lie on the cluster boundaries:

$$Z = \sum_{\mathcal{P}} \mathrm{q}^{\mathrm{B}}(1-\mathrm{q})^{\mathrm{W}} x^{\mathrm{length}} \,\mathrm{Tr} \prod_{\langle ij\rangle \in \partial\mathcal{P}} S_i^{\alpha} S_j^{\alpha}. \tag{5.8}$$

'Length' refers to the total length of cluster boundaries. Each link on a cluster boundary now carries a colour index α. Now integrate over $\mathbf{S}$ using

$$\mathrm{Tr}\, S_i^{\alpha} S_i^{\beta} = \delta^{\alpha\beta}, \quad \mathrm{Tr}\, S_i^{\alpha} S_i^{\beta} S_i^{\gamma} S_i^{\delta} = \frac{n}{n+2}\left(\delta^{\alpha\beta}\delta^{\gamma\delta} + \delta^{\alpha\gamma}\delta^{\beta\delta} + \delta^{\alpha\delta}\delta^{\beta\gamma}\right). \tag{5.9}$$

The three terms in the second formula correspond to the three ways of connecting up the links in pairs at a node for which all four links are in $\partial\mathcal{P}$. Expanding out all such brackets gives 3^N terms, each associated with a loop configuration $\mathcal{C}$, and each loop is consistently coloured. Summing over the colours,

$$Z = \sum_{\mathcal{C}} a^N \mathrm{q}^B (1-\mathrm{q})^W x^{\mathrm{length}} n^{\mathrm{no.\ loops}}. \tag{5.10}$$

The parameter a is $n/(n+2)$ as a consequence of Eq. 5.9; it can be varied by exchanging the hard constraint $\mathbf{S}^2 = n$ for a potential for $\mathbf{S}^2$.

5.3.2 Correlation Functions

As in Chap. 2 , the basic correlation functions are the watermelon correlators $G_k(x, y)$ (in a continuum notation). G_k may be written as the two point function of the k-leg operator O_k,

$$O_k(x) \propto S^1(x) S^2(x) \ldots S^k(x). \tag{5.11}$$

When inserted into a correlation function, this operator emits k strands (an even number) with colour indices ranging from 1 to k (Fig 5.6).

As usual, we will employ a replica limit to access the regime $n < 2$. As for the oriented loops, there is also a supersymmetric construction when n is an integer. The supervector $\mathbf{\Phi}_i$ which replaces $\mathbf{S}_i$ is of the form

$$\mathbf{\Phi}_i = (\mathbf{S}_i, \boldsymbol{\eta}_i, \boldsymbol{\xi}_i), \quad \mathbf{\Phi}_i^2 = n, \tag{5.12}$$

where $\mathbf{S}_i$ has $n+2m$ components and the fermions $\boldsymbol{\eta}_i, \boldsymbol{\xi}_i$ each have m anticommuting components. Inner products are defined by $\mathbf{\Phi}_i.\mathbf{\Phi}_j \equiv \mathbf{S}_i.\mathbf{S}_j + \boldsymbol{\eta}_i.\boldsymbol{\xi}_j + \boldsymbol{\eta}_j.\boldsymbol{\xi}_i$. The field $\mathbf{\Phi}$ lives on the supersphere $S^{n+2m-1|2m}$, but after identifying $\mathbf{\Phi}$ with $-\mathbf{\Phi}$ we have real projective superspace, $\mathrm{RP}^{n+2m-1|2m}$.

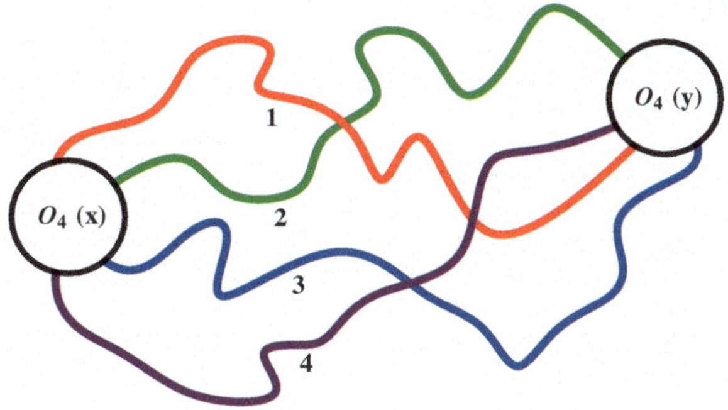

Fig. 5.6 The watermelon correlator $G_4(x, y)$

5.3.3 $\mathbb{Z}_2$ Vortices and $\mathbb{Z}_2$ Fluxes

We form the real symmetric matrix

$$Q^{\alpha\beta} = S^{\alpha} S^{\beta} - \delta^{\alpha\beta}, \quad \mathrm{tr}\, Q = 0, \tag{5.13}$$

which will be the relevant degree of freedom on long length scales, and which lives on RP^{n-1}. Since this manifold has nontrivial fundamental group [46], the Q configuration can have vortex defects which we now discuss.

RP^1 is equivalent to the circle, so at the special value $n = 2$ vortices are standard XY vortices and are characterized by an integer topological charge. However $\pi_1(\mathrm{RP}^{n-1}) = \mathbb{Z}_2$ when $n > 2$, so in general the vortex charge is defined only modulo two. In the replica limit—which requires analytic continuation of formulae defined for arbitrary integer n to $n < 2$—the vortices should again be viewed as $\mathbb{Z}_2$ vortices. (This may be clearer in the supersymmetric formulation of the $n = 1$ loop model, Eq. 5.12, where the bosonic part of the superspin lives on RP^{2m} for $m \geq 1$, which has fundamental group $\mathbb{Z}_2$.)

For a more concrete picture we return to $\mathbf{S}$ and σ. Let $\sigma_l = +1$ on all links l, except for a semi-infinite string of parallel links ending at a plaquette F (see Fig. 5.7). The flux $\prod \sigma$ is then -1 only on F. The spin configurations which maximize the Boltzmann weight vary smoothly except at the string, across which $\mathbf{S}$ changes sign. This indicates the presence of a $\mathbb{Z}_2$ vortex located at F. This is an example of a standard connection between vortices in nematics and $\mathbb{Z}_2$ fluxes [47].

Since vortices are associated with plaquettes of nontrivial flux, we can assign a vortex number to each plaquette which is $+1$ if $\prod \sigma = -1$ and 0 if $\prod \sigma = 1$. The gauge field part of the Boltzmann weight can then be written in terms of the number N_v of vortices:

$$\prod_F \left((1 - \mathrm{q}) + \mathrm{q} \prod_{\langle ij \rangle \in F} \sigma_{ij} \right) = (1 - 2\mathrm{q})^{N_v}. \tag{5.14}$$

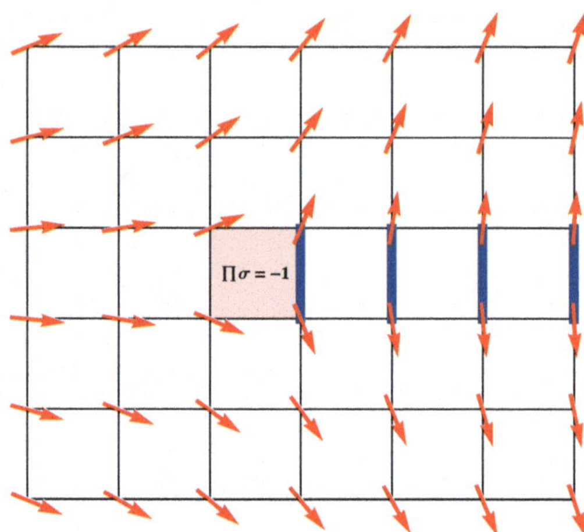

Fig. 5.7 Plaquettes with gauge flux $\prod \sigma = -1$ (shaded in *pink*) are endpoints of strings of links with $\sigma = -1$ (marked in *bold/blue*), across which $\mathbf{S}$ changes sign. In terms of the nematic order parameter, which is obtained by identifying $\mathbf{S}$ with $-\mathbf{S}$ and which lives on RP^{n-1}, these plaquettes are vortices (this is of course a caricature, neglecting fluctuations)

We see that the factor $(1-2\mathrm{q})$ is simply a fugacity for vortices, and that the exchange $\mathrm{q} \leftrightarrow 1-\mathrm{q}$ corresponds to changing the sign of the vortex fugacity.

This sign distinguishes the two short-loop phases in Fig. 5.5 from each other. In Ref. [20], $\mathbb{Z}_2$ vortices play an analogous role in a sigma model for localisation, with the sign of the vortex fugacity distinguishing two insulating phases. The vortex fugacity is also important for the critical behaviour (see Sect. 5.5.3).

Let us rewrite the Boltzmann weight for σ in the conventional form for $\mathbb{Z}_2$ gauge theory. In the absence of a boundary, Eq. 5.5 may be written

$$Z \propto \mathrm{Tr}\exp\Big(\kappa \sum_F \prod_{\langle ij\rangle \in F} \sigma_{ij}\Big) \prod_{\langle ij\rangle} \big(1 + x\, \sigma_{ij} \mathbf{S}_i.\mathbf{S}_j\big)\,, \tag{5.15}$$

where the gauge field stiffness is

$$\kappa = \frac{1}{2} \ln \frac{1}{|1-2\mathrm{q}|}. \tag{5.16}$$

In the presence of a boundary, denoted ∂, the Boltzmann weight acquires an additional term when $\mathrm{q} > 1/2$:

$$\prod_{\langle ij\rangle \in \partial} \sigma_{ij}. \tag{5.17}$$

This term effects the sign change in the vortex fugacity. It is equal to $(-)^{N_{\text{strings}}}$, where N_{strings} is the number of $\sigma = -1$ strings which terminate on the boundary. Since this number is equal to the number of vortices in the interior modulo 2, $(-)^{N_{\text{strings}}} = (-)^{N_v}$.

We see that the sign of the vortex fugacity does not affect bulk properties. Instead it determines the presence or absence of an edge loop.

Finally, consider the point q $= 1/2$. The vortex fugacity vanishes there (Eq. 5.14); however, the universal properties of this point do not differ from those in the rest of the Goldstone phase (Fig. 5.5). This is because vortices are anyway RG irrelevant in that phase. We discuss this in the next section in terms of a sigma model for Q.

The suppression of vortices (either microscopically or in the infrared) means that in the Goldstone phase this sigma model has a correspondence with the simpler $O(n)$ sigma model. In the IPLC at q $= 1/2$ this correspondence holds microscopically, since the gauge field stiffness diverges at this point (5.16). This enforces $\prod \sigma = 1$ for every face F, giving $\sigma_{ij} = \chi_i \chi_j$ (so long as the lattice lives on a simply-connected manifold[3])

$$Z \propto \sum_{\{\chi\}} \mathrm{Tr}_S \prod_{\langle ij \rangle} \left(1 + x\,(\chi_i \mathbf{S}_i).(\chi_j \mathbf{S}_j)\right). \tag{5.18}$$

Changing variables to $\mathbf{S}' = \chi \mathbf{S}$ eliminates the gauge degrees of freedom from the Boltzmann weight, leaving a lattice $O(n)$ model. (Non-gauge-invariant correlators pick up factors of χ, ensuring that they vanish[4] on summing over χ.) In the regime we consider, i.e. at sufficiently large[5] x, this $O(n)$ model is expected to be described by the $O(n)$ sigma model in its Goldstone phase.

5.3.4 Continuum Description

The naive continuum description of the lattice field theory (5.5) is the sigma model Lagrangian

$$\mathcal{L} = \frac{K}{4}\,\mathrm{tr}\,(\nabla Q)^2, \tag{5.19}$$

together with the constraint on Q following from its definition in terms of $\mathbf{S}$. To conform with convention we use the normalisation $\mathbf{S}^2 = 1$ in the continuum; then $Q^{\alpha\beta} = S^\alpha S^\beta - \frac{1}{n}\delta^{\alpha\beta}$.

The fugacity for vortices is hidden in the ultraviolet regularisation of (5.19). We will restore this parameter explicitly when we consider RG in the vicinity of the critical point (Sect. 5.5.3), where vortices are crucial.

However, the classical free energy of a pair of vortex defects is proportional to the stiffness K, as in the XY model, so vortices are suppressed at large K. In the

---

[3] Otherwise there can be flux through holes or handles of the manifold. In the graphical expansion, the sum over the resulting flux sectors is responsible for killing loop configurations which do not correspond to percolation configurations.

[4] However, at this value of q we can define k-leg operators for odd k by taking $O_k = S'^1 \ldots S'^k$.

[5] Locations $x_c(n)$ of the dilute critical point in the square lattice $O(n)$ model are determined numerically in a recent paper [48].

Goldstone phase, which we now discuss, K flows to large values under coarse-graining, and vortices are an irrelevant perturbation.

Non-singular (vortex-free) configurations of Q are equivalent to non-singular configurations of $\mathbf{S}$ (on a manifold of trivial topology, and up to a global sign ambiguity). Thus for a perturbative treatment at large K, the RP^{n-1} sigma model can be replaced with the more familiar $O(n)$ sigma model, argued previously [11] to apply to the CPLC at $q = 1/2$:

$$\mathcal{L} = \frac{K}{2}(\nabla \mathbf{S})^2, \quad \mathbf{S}^2 = 1. \tag{5.20}$$

The perturbative beta function for K changes sign at $n = 2$ [11, 49]. To two-loop order,

$$\frac{\mathrm{d}K}{\mathrm{d}\ln L} = \frac{2-n}{2\pi}\left(1 + \frac{1}{2\pi K} + \cdots\right) \tag{5.21}$$

When $n > 2$, the stiffness flows to zero under RG and the sigma model has only a disordered phase; thus the loop model is not expected to be critical. At $n = 2$ the sigma model is the XY model, and we have in addition the quasi-long-range-ordered phase in which K does not flow. By the Mermin Wagner theorem, these are the only possibilities when $n \geq 2$.

However in the replica limit the Mermin Wagner theorem does not apply [11], and Eq. 5.21 shows that for $n < 2$ the stiffness K flows to infinity in the infrared. This is the Goldstone phase. At the infra-red fixed point the $(n-1)$ Goldstone modes are free fields, so the central charge is $c = n - 1$ [11].

We now discuss the extraction of a continuum description for the CPLC, the appearance of a nontrivial vortex fugacity in that model, and the extra symmetry on the boundaries of the phase diagram.

5.3.5 Lattice Field Theory for the CPLC

The CPLC can again be mapped to a lattice spin model with a $\mathbb{Z}_2$ gauge symmetry, though with a less conventional form, and as long as we are in the interior of the phase diagram (Fig. 5.3) the continuum description is again the RP^{n-1} model.

The required construction is an immediate extension of that for the models in Chap. 2. We place spins $\mathbf{S}_l$ ($\mathbf{S}_l^2 = n$) on the links l of the square lattice, with

$$Z = \mathrm{Tr}\,\exp\left(-\sum_{\text{nodes}} S_{\text{node}}\right). \tag{5.22}$$

Denoting the links surrounding i by $1, 2, 3, 4$, with the weight p pairing being 1 with 3 and 2 with 4, and the weight $(1-p)q$ pairing being 1 with 2 and 3 with 4,

$$\exp(-S_{\text{node}}) = p(\mathbf{S}_1.\mathbf{S}_3)(\mathbf{S}_2.\mathbf{S}_4) + (1-p)q(\mathbf{S}_1.\mathbf{S}_2)(\mathbf{S}_3.\mathbf{S}_4) \\ + (1-p)(1-q)(\mathbf{S}_1.\mathbf{S}_4)(\mathbf{S}_2.\mathbf{S}_3).$$

A graphical expansion gives the CPLC in the usual way. The above Boltzmann weight again has a $\mathbb{Z}_2$ gauge symmetry: on changing the sign of the spin on the link ij, both e^{-S_i} and e^{-S_j} change sign but the overall Boltzmann weight is unchanged. The naive continuum limit is the RP^{n-1} sigma model described above, and the universal properties are expected to be identical with those of the IPLC. In Sect. 5.3.7 we will argue that the effective fugacity for $\mathbb{Z}_2$ vortices changes sign on the line $q = 1/2$, just as for the IPLC. Note that, as for the models of Chap. 2, the lattice Boltzmann weight is not necessarily positive.

5.3.6 Phase Diagram Boundaries and $\mathbf{CP}^{n-1}$

On the boundaries of the phase diagram (when $p = 0$, or $q = 0$, or $q = 1$) there is a protocol for orienting the loops, as mentioned in Sect. 5.2.1. Equation 5.22 may be replaced by a lattice CP^{n-1} model (Chap. 2), showing that the $SO(n)$ global and $\mathbb{Z}_2$ gauge symmetry of the interior of the CPLC phase diagram are promoted to an $SU(n)$ global and $U(1)$ gauge symmetry on its boundary.

The CP^{n-1} Lagrangian, including the θ term, was displayed in Eq. 3.1. In this chapter we will denote the matrix CP^{n-1} field by $\widetilde{Q}$ to distinguish it from the real-valued Q above.

The CP^{n-1} description tells us that the left- and right-hand boundaries of the phase diagram Fig. 5.3 (the Manhattan lattice loop model) are localized for all p as previously expected [26, 41]. Here θ can be shown[6] to be equal to zero mod 2π, so that the sigma model is in the disordered phase. However the correlation length ξ, and the typical loop size, diverge exponentially as $p \to 1$. For $n = 1$,

$$\xi \sim (1-p)^{-2} e^{\text{const.}/(1-p)} \qquad \text{(when } q = 0, 1\text{)}.$$

This follows from the beta function for the CP^{n-1} model [51] and the fact that the bare stiffness is of order $(1-p)^{-1}$. It is the behaviour of ξ for non-critical localisation in class C [37, 38], to which the Manhattan lattice loop model is related [26].

The CP^{n-1} description of the model on the line $p = 0$ gives a way of seeing that the vortex fugacity in the RP^{n-1} description of the CPLC changes sign on the central line $q = 1/2$, just as it does at the central point q $= 1/2$ in the IPLC (though this can also be seen directly in the lattice RP^{n-1} model). We now discuss this.

[6] E.g. by mapping an anisotropic version of the loop model to a spin chain with next-nearest-neighbour couplings of the kind in Ref. [50], or by coarse graining the lattice CP^{n-1} model.

5.3.7 Vortex Fugacity and the θ-term for CP^{n-1}

A natural way to approach the field theory for the CPLC, at least when the weight p for crossings is small, is by perturbing the field theory for the model without crossings. We may check in the lattice CP^{n-1} model that the perturbation corresponds to a mass for the imaginary part of $\widetilde{Q}$, i.e. $\delta\mathcal{L} \propto -p\,\mathrm{tr}\,(\mathrm{Im}\,\widetilde{Q})^2$. As in Chap. 4, we expect this to lead in the infrared to an RP^{n-1} model.

If we simply set $\widetilde{Q}$ to be real, the kinetic term for $\mathcal{L}_{\mathrm{CP}^{n-1}}$ becomes that of $\mathcal{L}_{\mathrm{RP}^{n-1}}$ (5.19), and the θ term vanishes. However, this does not mean the θ term plays no role: it vanishes only if we neglect RP^{n-1} vortices. In the presence of a vortex, we must allow the imaginary part of $\widetilde{Q}$ to become nonzero in the vortex core, in order to retain continuity of $\widetilde{Q}$ and the sigma model constraint. There is then a contribution from the θ term there. This mechanism also pertains to quantum magnets described by anisotropic sigma models in both $1+1$ and $2+1$ dimensions [52, 53].

The vortex core corresponds either to a half-skyrmion or to an anti-half-skyrmion, depending on the sign of the imaginary components of $\widetilde{Q}$. This can easily be visualised for $n = 2$, when the perturbed CP^1 model is equivalent to the $O(3)$ sigma model with easy-plane anisotropy: a half-skyrmion corresponds to a vortex in the easy plane, with the $O(3)$ spin perpendicular to the easy plane in the core. The effective vortex fugacity V is obtained by summing over the two possibilities, half-skyrmion and anti-half-skyrmion, leading (by the relation between the θ term and the topological density, Sect. 3.2.1) to

$$V \propto e^{i\theta/2} + e^{-i\theta/2}.$$

Since at $p = 0$ we have $(\theta - \pi) \sim (q - 1/2)$, the vortex fugacity in the RP^{n-1} description of the CPLC changes sign at $q = 1/2$, just as the vortex fugacity changes sign at q $= 1/2$ in the IPLC.

We can also see this directly in the lattice RP^{n-1} model (5.22) without appeal to the CP^{n-1} model, as described in Appendix B. The mechanism is similar to that discussed for hedgehogs on the 3D L lattice in Chap. 3.

5.4 The Goldstone Phase

The Goldstone phase shows subtle universal behaviour, which is very different from that seen in loop models without crossings but which can nevertheless be understood in detail. Within this phase we can work with the $O(n)$ sigma model [11], as discussed in Sect. 5.3.4.

For most purposes the Lagrangian (5.20) will be sufficient, but to calculate the length distribution in Sect. 5.4.3 we will need to add a small perturbation, γ, which breaks the symmetry from $O(n)$ to $O(n-1) \times \mathbb{Z}_2$. Writing

$$\mathbf{S} = (S^1, \mathbf{S}_\perp), \tag{5.23}$$

where $\mathbf{S}_\perp$ is an $(n-1)$-component vector, we take

$$\mathcal{L} = \frac{K}{2}\left((\nabla\mathbf{S})^2 + \gamma\,\mathbf{S}_\perp^2\right), \quad \mathbf{S}^2 = 1. \tag{5.24}$$

We briefly recall one-loop RG results for this model [49, 54, 55], which may be obtained easily using the background field method. $\mathbf{S}(x)$ is decomposed into a slowly-varying field $\tilde{\mathbf{S}}(x)$ and rapidly varying fluctuations $\phi(x)$:

$$\mathbf{S}(x) = \tilde{\mathbf{S}}(x)\sqrt{1-\phi(x)^2} + \sum_{a=1}^{n-1}\phi_a(x)\mathbf{e}_a(x). \tag{5.25}$$

The $e_a(x)$, $a = 1, \ldots, n-1$, are a set of vectors orthogonal to $\tilde{\mathbf{S}}(x)$ (there is a gauge freedom in this choice). If the initial UV cutoff is Λ, so that $\mathbf{S}$ involves modes with wavenumber $|k| < \Lambda$, then the modes in $\tilde{\mathbf{S}}(x)$ are limited to $|k| < \tilde{\Lambda}$ for the new cutoff $\tilde{\Lambda} < \Lambda$, and ϕ contains modes in the momentum shell $|k| \in (\tilde{\Lambda}, \Lambda)$. Integrating ϕ out, and working to leading order in K^{-1} and γ, we obtain the RG equations

$$\frac{\mathrm{d}K}{\mathrm{d}\tau} = \frac{2-n}{2\pi}, \quad \frac{\mathrm{d}\gamma}{\mathrm{d}\tau} = \left(2 - \frac{1}{\pi K}\right)\gamma. \tag{5.26}$$

Here τ is the RG time: after time τ, the new cutoff is $e^{-\tau}\Lambda$. Again, the important point is that K flows to large values in the infrared if $n < 2$ [11].

Note that in two dimensions the higher-order anisotropies (higher powers of $\mathbf{S}_\perp^2$) are as relevant as $\mathbf{S}_\perp^2$ at tree level, but less relevant at one-loop order. They will be important for the polymer phase diagram discussed in Chap. 6.

We now calculate a range of observables both analytically and numerically; details of the numerical procedure are given in Sect. 5.6. As we discuss in Chap. 6, incompatible hypotheses about the universal behaviour at the point $n = 1$, $p = 1/3$, $q = 1/2$ have previously been put forward, which is one reason for making a careful comparison of numerics and theory in the Goldstone phase.

5.4.1 Correlation Functions

The watermelon correlation function $G_k(r)$ is the two-point function of the operator $S^1, \ldots, S^k(x)$ (Sect. 5.3.2). Including the UV cutoff Λ and the sigma model stiffness K explicitly in the argument of G_k, a simple calculation following [49] gives

$$G_k(r\Lambda, K) = \frac{G_k\left(1, K + \frac{2-n}{2\pi}\ln\Lambda r\right)}{\left(1 + \frac{2-n}{2\pi K}\ln\Lambda r\right)^{\xi_k}}, \tag{5.27}$$

where the exponent in the denominator depends on k and on the loop fugacity n:

$$\xi_k = \frac{k\,(k+n-2)}{2-n}. \tag{5.28}$$

The correlation function in the numerator of (5.27), which is evaluated at a separation of the order of the new UV cutoff, tends to a constant for large r. Thus the asymptotic behaviour of the watermelon correlation functions $G_k(r)$ is given by a universal power of $\log r$:

$$G_k(r) = \frac{C_k}{(\ln r/r_0)^{\xi_k}}, \qquad r_0 = \Lambda^{-1} e^{-\frac{2\pi K}{2-n}}, \tag{5.29}$$

with nonuniversal C_k, r_0.

It is interesting to note that although the stiffness K flows to infinity in the Goldstone phase—which we would usually think of as implying long range order—*all* the correlation functions G_{2l} decay to zero at large distances for $n > 0$. Correlators G_k with odd k have no meaning in the RP^{n-1} sigma model or in the CPLC, but can be defined in loop models described by the $O(n)$ sigma model without a $\mathbb{Z}_2$ gauge symmetry; such models allow operators which insert dangling ends. For $n = 1$, G_1 tends to a constant, indicating that the entropic force between the ends of an *open* strand inserted into such a soup of closed loops vanishes at large separations.

We may contrast this with the case $n = 0$, which describes the universality class of the dense polymer with crossings [11]—a single loop whose length is comparable with the number of lattice sites. Here, $G_2(x)$ is a constant at large separations, in consequence of this fact. On the other hand $G_1(x)$ has a *negative* exponent ξ_1, indicating that if the polymer is an open strand the two ends suffer a weak entropic repulsion.

We might have expected the logarithmic form of (5.29) to prevent us from seeing the universal exponents ξ_k numerically, but this is not the case. Figure 5.8 shows

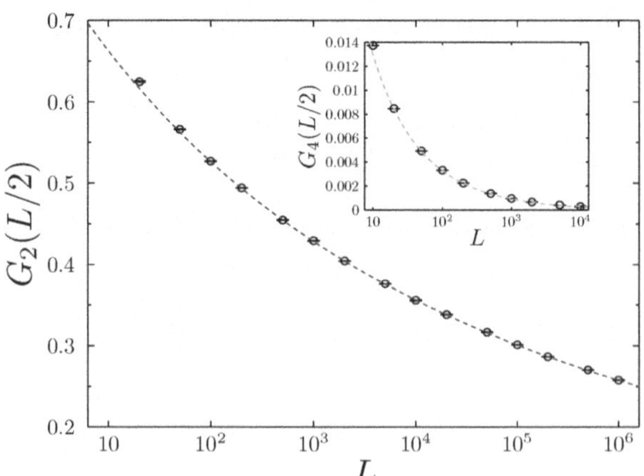

Fig. 5.8 Two- and four-leg watermelon correlation functions G_2 and G_4 in the Goldstone phase. The fits are to the form $G_k = \tilde{C}_k\,(\ln L/r_k)^{-\tilde{\xi}_k}$ $(k = 2, 4)$, see text

$G_2(L/2)$ and $G_4(L/2)$ for $L \times L$ systems with periodic boundary conditions, with L ranging up to $L = 10^6$ for G_2 and $L = 10^4$ for G_4. Simulations are at $p = q = 1/2$. We fit G_2 and G_4 to the form $G_k = \tilde{C}_k (\ln L/r_k)^{-\tilde{\xi}_k}$, leading to exponents consistent with (5.28):

$$\tilde{\xi}_2 = 1.9(1), \quad \tilde{\xi}_4 = 12.5(10). \tag{5.30}$$

We have $\ln r_2 = -15.4(14)$, $\ln r_4 = -18(2)$, consistent with the fact that r_0 is shared between different G_k in Eq. 5.29.

5.4.2 Spanning Number

The logarithmic RG flow of the sigma model stiffness K can be seen empirically: this stiffness is directly related to the mean spanning number for an $L \times L$ cylinder on which curves are allowed to terminate on the boundary. This is the number n_s of curves which traverse the cylinder from one boundary to the other. Note that n_s must be even if L is even and odd if L is odd.

To calculate n_s, the correspondence of Sect. 5.3.5 between the loop model and a spin model must be extended to the case with dangling boundary links. We take the spins on the dangling links to be fixed, with

$$\frac{\mathbf{S}_{\text{top}}}{\sqrt{n}} = (\cos\theta, \sin\theta, 0, \ldots, 0), \qquad \frac{\mathbf{S}_{\text{bottom}}}{\sqrt{n}} = (1, 0, \ldots, 0)$$

on the top and bottom boundaries (above we temporarily revert to the lattice normalization of $\mathbf{S}$). The graphical expansion then goes through as before, except that spanning curves acquire an additional weight $\cos\theta$.[7]

Denoting the partition function with the above boundary conditions by $Z(\theta)$, we therefore have

$$\left\langle (\cos\theta)^{n_s} \right\rangle = \frac{Z(\theta)}{Z(0)}. \tag{5.31}$$

In the Goldstone phase the stiffness flows to large values in the infra-red, so to calculate the right hand side we need consider only classical solutions with the appropriate boundary conditions. Letting x be the coordinate along the cylinder, these are

$$\mathbf{S} = \pm(\cos\phi(x), \sin\phi(x), 0, \ldots, 0), \qquad \phi(x) = \frac{x(\theta + \pi m)}{L}.$$

[7] Loops in the interior and curves with both ends on the same boundary retain fugacity n—for example a curve with both ends on the upper boundary can either be of colour $\alpha = 1$, in which case it has weight $n \cos^2\theta$, or of colour $\alpha = 2$, when it has weight $n \sin^2\theta$, and the sum gives n. Spanning curves on the other hand must be of colour $\alpha = 1$, and have weight $n \cos\theta$.

Both odd and even values of m are allowed—the boundary condition is satisfied only up to a sign—but when L and m are both odd the Boltzmann weight acquires an additional minus sign, as can be seen from the lattice partition function.[8]

The action of these solutions is calculated using the renormalized stiffness $\widetilde{K}$ on scale L, leading to

$$\left\langle (\cos\theta)^{n_s} \right\rangle \simeq \sum_m (-)^{mL} \exp\left(-\frac{\widetilde{K}}{2}(\theta - \pi m)^2 \right). \tag{5.32}$$

For a given value of θ, only one or two values of m are not exponentially small in $\widetilde{K}$.

To extract low-order cumulants for n_s, we set $\cos\theta = e^{-x}$ and expand in x. Since (5.32) is dominated by the $m = 0$ term for $\theta \sim 0$, the difference between even and odd L is not seen. The lth cumulant is given by

$$\left\langle\!\left\langle n_s^l \right\rangle\!\right\rangle \simeq -\frac{\widetilde{K}}{2} \partial_y^l (\arccos e^y)^2 \mid_{y=0}.$$

In particular, the mean spanning number is given by the renormalized stiffness $\widetilde{K}$, so

$$\langle n_s \rangle \sim \frac{2-n}{2\pi} \ln \frac{L}{L_0}. \tag{5.33}$$

This logarithmic flow (for $n = 1$) is seen in Fig. 5.9 for two points in the Goldstone phase. We have fitted the data for large sizes to the slightly more accurate form $\langle n_s \rangle \simeq \frac{1}{2\pi}(\ln L/L_0 + \ln\ln L/L_0)$ which comes from including the subleading $O(1/K)$ term in the beta function for the stiffness (5.21). In the upper inset to Fig. 5.9 we plot the numerical value of the slope $\mathrm{d}\langle n_s \rangle / \mathrm{d}\ln L$, which is seen to converge slowly to $1/2\pi$ for large L.

Since all cumulants are proportional to $\widetilde{K}$, their ratios are universal numbers which we can compare with data:

$$\left\langle\!\left\langle n_s^2 \right\rangle\!\right\rangle = \frac{2}{3}\langle n_s \rangle, \qquad \left\langle\!\left\langle n_s^3 \right\rangle\!\right\rangle = \frac{4}{15}\langle n_s \rangle.$$

These relations are obeyed to good accuracy—plotting the two cumulants above against $\langle n_s \rangle$ for $p = 1/2$ and various L gives straight lines with slopes 0.668(5) and 0.274(18) respectively (data not shown). Note that the scaling of the cumulants means that when $\langle n_s \rangle$ becomes very large the probability distribution P_{n_s} for the spanning number becomes Gaussian (away from its tails).

To extract the probability distribution for small integer n_s we set $\cos\theta = \epsilon$ in (5.32),

$$\sum_{n_s} P_{n_s} \epsilon^{n_s} \simeq e^{-\frac{\widetilde{K}}{2}(\arccos\epsilon)^2} + (-)^L e^{-\frac{\widetilde{K}}{2}(\arccos\epsilon - \pi)^2}. \tag{5.34}$$

[8] In the mapping of Sect. 5.3.5 this sign comes from factors of $\mathbf{S}_{\text{top}}.\mathbf{S}_l$ and $\mathbf{S}_{\text{bottom}}.\mathbf{S}_l$ in the Boltzmann weight for links l adjacent to the boundary links.

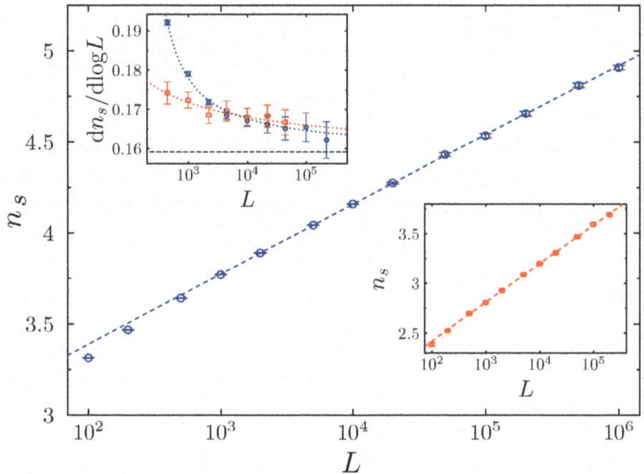

Fig. 5.9 The logarithmic increase of the mean spanning number $\langle n_s \rangle$ with system size in the Goldstone phase. *Main panel* $p = 1/2$, $q = 1/2$; *lower inset* $p = 1/3$, $q = 1/2$. Fits are to $\frac{1}{2\pi}(\ln L/L_0 + \ln\ln L/L_0)$ with $\ln L_0 \simeq -13.78, -8.06$ for $p = 1/2$ and $p = 1/3$ respectively. *Upper inset* shows the numerical estimates for the slopes $\mathrm{d}\langle n_s \rangle / \mathrm{d}\ln L$ plotted against $\ln L$—they are expected to converge to $1/2\pi$ (*horizontal line*)

For even circumference, this gives for example

$$P_0 = 2e^{-\pi^2\langle n_s \rangle/8}, \quad P_2 = \frac{\pi^2 \langle n_s \rangle^2 - 4\langle n_s \rangle}{4} e^{-\pi^2\langle n_s \rangle/8}.$$

In Fig. 5.10 the expressions for $P_0, \ldots, P_8$ are compared with data (at $p = q = 1/2$ and L in the range 10^2–10^6) showing remarkable agreement. There is no free parameter in these fits.

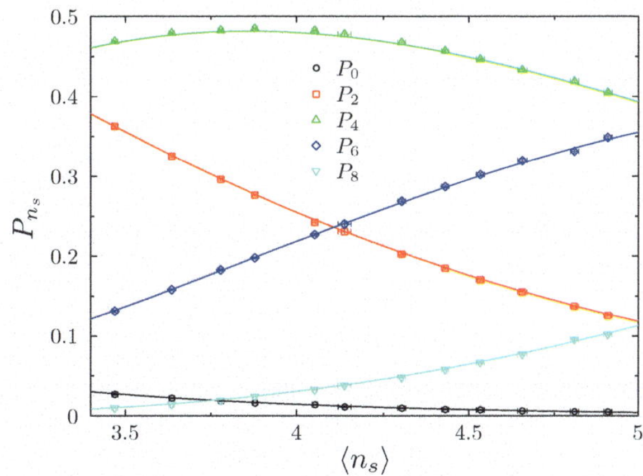

Fig. 5.10 Probabilities P_{n_s} of n_s spanning curves, as a function of $\langle n_s \rangle$ (data for $p = q = 1/2$). *Curves* are the analytical expressions from Eq. 5.34, with $\widetilde{K} = \langle n_s \rangle$

5.4.3 Length Distribution

To calculate the length distribution for a loop we must consider the RG flow away from the Goldstone phase induced by a symmetry-breaking perturbation. To summarize the result of the following calculation, which is for $n = 1$, the probability for a loop randomly chosen from the soup to have length l falls off as

$$P(l) \propto \frac{1}{l^2 \ln^2(l/l_0)} \tag{5.35}$$

at large l.

Figure 5.11 shows the distribution obtained numerically for loops of length up to $l \sim 10^{10}$. We multiply $P(l)$ by l^2 in order to expose the logarithmic correction, which we fit to the form $a(\ln l/l_0)^{-c}$. We obtain

$$c = 2.03(3) \tag{5.36}$$

in striking agreement with (5.35). This value is also in agreement with numerical results for trails [16, 18], as we will discuss shortly.

Note that $P(l)$ differs by a factor of l from the length distribution for the loop passing through a fixed link,

$$P_{\text{fixed link}}(l) \propto l\, P(l), \tag{5.37}$$

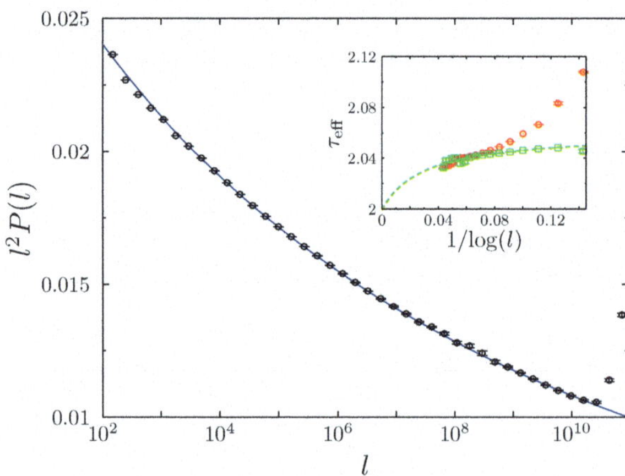

Fig. 5.11 *Main panel* The probability distribution $P(l)$ for the length of a loop in the Goldstone phase. We multiply by l^2 to remove the expected power law part, leaving the logarithmic dependence. The fit is as described in the text, with $\ln l_0 = -33.7(8)$. Data is for $p = q = 1/2$. *Inset* two ways of defining the effective finite-size exponent value τ_{eff} (see text). *Green squares* data from $P(l)$, together with the fit (*dashed line*) implied by Eq. 5.35 ($\ln l_0 = -32.7$). *Red circles* data from $\Delta X(l)$

simply because longer loops visit more links. Thus $\langle l \rangle$ evaluated using $P(l)$ is finite, as for (5.35).

Let $g(x)$ denote the generating function for the length of a loop randomly chosen from the soup (the 'marked' loop):

$$g(x) = \sum_l P(l)x^l = \left\langle x^{\text{length of marked loop}} \right\rangle. \tag{5.38}$$

In order to extract $g(x)$ we use the trick of Ref. [56], splitting the components of $\mathbf{S}$, or equivalently the loop colours, into two groups. For simplicity we will consider only the loop model at fugacity one, though it would be easy to generalize. We split $\mathbf{S}$ as in Eq. 5.23, yielding an $(n-1)$-component vector $\mathbf{S}_\perp$. In the graphical expansion of a lattice model, say of the CPLC (the IPLC would be similar) we correspondingly split the loops into unmarked loops, whose colour index is equal to one, and marked loops, whose colour index runs over $2, \ldots, n$. After summing over loop colours, a configuration with N marked loops has a weight $(n-1)^N$, and expanding the partition function in $(n-1)$ is equivalent to expanding in the number of marked loops in the configuration. Writing $n' = n - 1$,

$$Z(n') = \sum_{\mathcal{C}} W_{\mathcal{C}} + n' \sum_{\substack{\mathcal{C};\ \text{one} \\ \text{marked loop}}} W_{\mathcal{C}} + \cdots \tag{5.39}$$

Here $W_{\mathcal{C}}$ is the weight of a configuration $\mathcal{C}$ in the CPLC at $n = 1$ (Eq. 5.1). The first term on the right hand side is equal to one.

Next, we wish to modify the weight of a configuration by the factor x^l, where l is the total length of marked loops. This is easily done: all the inner products $\mathbf{S}_l.\mathbf{S}_{l'}$ appearing in the node factors e^{-S_i} (see Sect. 5.3.5) are replaced according to:

$$\mathbf{S}_l.\mathbf{S}_{l'} \quad \longrightarrow \quad S_l^1 S_{l'}^1 + x\, \mathbf{S}_{l\perp}.\mathbf{S}_{l'\perp}. \tag{5.40}$$

In this way every unit length of marked loop picks up a factor of x, and the graphical expansion yields

$$Z(n', x) = 1 + n' \sum_{\substack{\mathcal{C};\ \text{one} \\ \text{marked loop}}} W_{\mathcal{C}}\, x^l + \cdots \tag{5.41}$$

Differentiating with respect to n' gives the required generating function:

$$g(x) = \frac{\partial_{n'} Z(n', x)\big|_{n'=0}}{\partial_{n'} Z(n', 1)\big|_{n'=0}}. \tag{5.42}$$

In terms of the free energy density $f(n', x)$,

$$Z(n', x) = e^{-L^2 f(n',x)}, \qquad f(n', x) = f_0 + n' f_1(x) + \cdots,$$

this is $g(x) = f_1(x)/f_1(1)$. In order to calculate the length distribution for large values of l, we require the free energy for $x \simeq 1$ and to first order in n'.

In the continuum description, the symmetry-breaking perturbation $x \neq 1$ leads to an infinite number of relevant perturbations of which the most relevant is that in Eq. 5.24. Setting $x = \exp(-\mu)$ with $\mu \ll 1$, we have $\gamma \sim \mu$, the constant of proportionality being nonuniversal.

Beginning with the Lagrangian of Eq. 5.24, we integrate out high frequency modes, retaining their contribution to the free energy, up to an RG time τ_*. This gives

$$f(K, \gamma) = f^<(K, \gamma) + f^>(K, \gamma), \tag{5.43}$$

where we have split up the contribution from the modes that have been integrated out,

$$f^<(K, \gamma) = \frac{n'}{4\pi} \int_0^{\tau_*} \frac{d\tau}{e^{2\tau}} (\ln K(\tau) + \gamma(\tau)) - n'A$$

(the nonuniversal constant A ensures $f < 0$, as required by Eq. 5.41) and those remaining:

$$f^>(K, \gamma) = e^{-2\tau_*} f(K(\tau_*), \gamma(\tau_*)).$$

The solutions to the RG equations (5.26) are

$$K(\tau) = K + \frac{\tau}{2\pi}, \qquad \gamma(\tau) = \gamma e^{2\tau} \left(\frac{2\pi K}{2\pi K + \tau} \right)^2. \tag{5.44}$$

Stopping the RG when γ_* becomes of order one, $f^>$ may be approximated as the free energy of a massive theory in which $\mathbf{S}$ executes only small fluctuations around $\mathbf{S} = (1, 0, \ldots, 0)$:

$$\mathcal{L} = \frac{K_*}{2} \left((\nabla \mathbf{S}_\perp)^2 + \gamma_* \mathbf{S}_\perp^2 + O(\mathbf{S}_\perp^4, \mathbf{S}_\perp^2/K_*) \right). \tag{5.45}$$

(The $O(\mathbf{S}_\perp^2/K_*)$ term comes from the sigma model measure.) However, the dominant terms in f come from $f^<$. To the order that we require,

$$f(K, \gamma) = -B + \gamma \left(2\pi K - \frac{8\pi^2 K^2}{\ln 1/\gamma} + \cdots \right) \tag{5.46}$$

(B is a constant.) We thus have the form of the generating function at small μ:

$$\left\langle e^{-\mu l}\right\rangle = 1 - C\,\mu\left(1 - \frac{4\pi K}{\ln 1/\mu} + \cdots\right). \tag{5.47}$$

The constants C and K are non-universal, and the fact that the leading μ dependence is linear in μ is simply a consequence of the fact that $\langle l\rangle$ is finite. However we may infer the behaviour of $P(l)$ at large l from the form of the nonanalytic term, yielding Eq. 5.35.

Previous work on self-avoiding trails [16, 18], which map to the $n = 1$ loop model at $p = 1/3$, $q = 1/2$ (see Chap. 6), considered a probability $Q(l)$ which may be written

$$Q(l) = \int_l^\infty P_{\text{fixed link}}(l')\mathrm{d}l'. \tag{5.48}$$

Viewing the loops as the trajectories of walkers, $Q(l)$ is the probability that a walker has not yet returned to its starting point after l steps. Equations 5.35, 5.37 and 5.48 give $Q(l) \sim 1/\ln l$. This agrees with the scaling found numerically in Refs. [16, 18].

For a generic critical loop ensemble, $P(l) \sim l^{-\tau}$ for some $\tau \geq 2$, and the mean size ΔX of a loop scales with its length as $\Delta X \sim l^{1/d_f}$. The fractal dimension d_f is related to τ by

$$\tau = 2/d_f + 1. \tag{5.49}$$

In the Goldstone phase, $\tau = d_{\text{eff}} = 2$, with logarithmic corrections. We may define finite size estimates of τ either using $\mathrm{d}\ln P/\mathrm{d}\ln l$ or using $\mathrm{d}\ln \Delta X/\mathrm{d}\ln l$ and the scaling relation. These are plotted in the inset to Fig. 5.11; both are expected to converge to two, but with different logarithmic corrections.

Here, ΔX is defined as the mean extent of a loop in one of the coordinate directions. A similar quantity—the mean square end to end distance of an open trail in the ISAT model—was considered numerically by Owczarek and Prellberg [16], and logarithmic corrections to Brownian scaling were found. It would be interesting to calculate these quantities analytically.

5.5 The Critical Lines

The critical lines separate the Goldstone phase from phases with short loops. In the language of the RP^{n-1} model, they correspond to order-disorder transitions at which $\mathbb{Z}_2$ vortices are set free. In this section we give numerically determined critical exponents for this transition at $n = 1$ and briefly consider an approximate RG treatment of vortices [19, 20, 57].

5.5.1 Critical Spanning Number, ν, and y_{irr}

At the critical point, the dimensionless quantity n_s (defined in Sect. 5.4.2) is expected to take a universal value. This is manifested in the crossings of the various curves in Fig. 5.12, which shows the spanning number n_s as a function of q for $p = 1/2$ and for cylinders of various sizes. Figure 5.13 shows the same quantity, but in the immediate vicinity of the critical point and including much larger system sizes (up to $L = 128{,}000$). The main panel of Fig. 5.14 shows data for $p = 0.3$; here finite size effects—visible in the drift in crossings—are much stronger.

The basic finite size scaling form for n_s is

$$n_s = h(x), \qquad x = L^{1/\nu}\delta q, \tag{5.50}$$

where $\delta q = q - q_c$. We take into account also nonlinear dependence of the scaling variable x on δq, replacing the second equation above with

$$x = L^{1/\nu}\delta q\left(1 + \beta_1\delta q + \beta_2\delta q^2\right), \tag{5.51}$$

and finite size corrections with (negative) irrelevant exponent $y_{\rm irr}$ in the form:

$$n_s = h(x)\left(1 + L^{y_{\rm irr}}(\beta_3 + \beta_4 x)\right). \tag{5.52}$$

A reasonable scaling collapse may be obtained by adjusting the values of q_c, β_i, ν and $y_{\rm irr}$. To find these values we fit n_s to the form (5.52), constructing $h(x)$ using B-splines with 22 degrees of freedom. The result for $p = 1/2$ is shown in the inset to Fig. 5.12. What is plotted is $n_s^F = n_s/\left(1 + L^{y_{\rm irr}}(\beta_3 + \beta_4 x)\right)$, which should be equal to the scaling function $h(x)$ by (5.52).

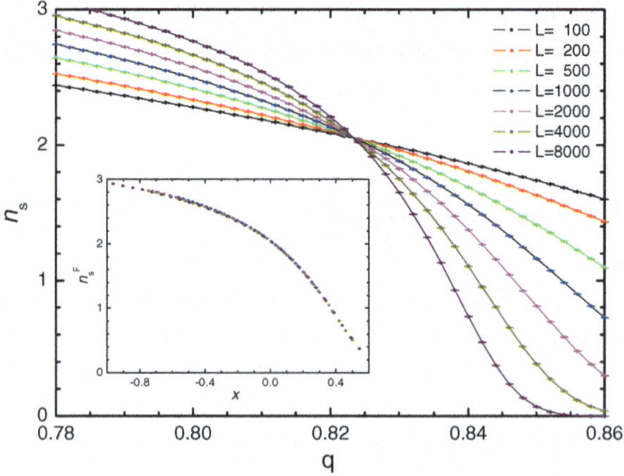

Fig. 5.12 *Main panel* the mean spanning number n_s as a function of q for $p = 1/2$ and various system sizes, showing a crossing at the transition. *Inset* data collapsed according to Eqs. 5.51 and 5.52

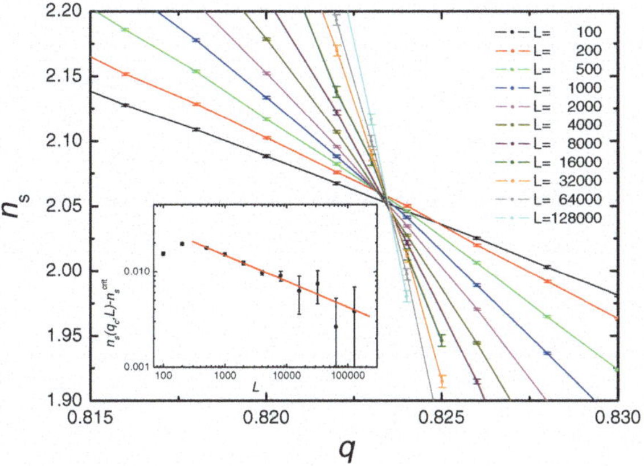

Fig. 5.13 *Main panel* the behaviour of the mean spanning number at $p = 1/2$ very close to the critical point. Note the larger system sizes compared to Fig. 5.12. *Inset* finite size corrections to the spanning number at the critical point (note log-log scale) and linear fit leading to estimate of y_{irr}

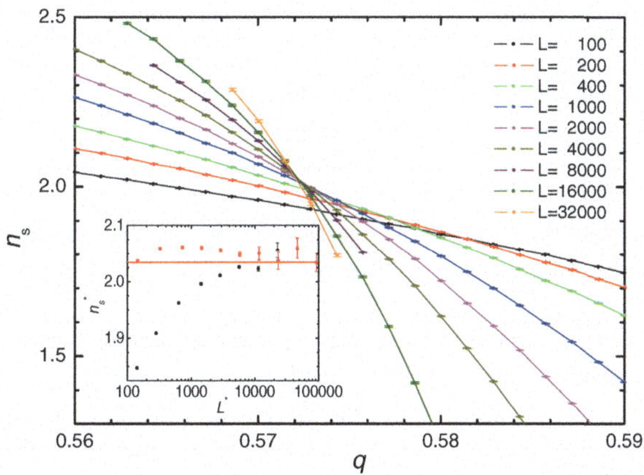

Fig. 5.14 *Main panel* spanning number at $p = 0.3$ (close-up of the critical point) showing larger drift in crossing points than at $p = 1/2$ (however, note different scale). *Inset* the vertical coordinate n_s^* is the crossing in the spanning number between consecutive system sizes and L^* is the geometrical mean of the sizes. *Red* and *black dots* correspond to $p = 1/2$ and $p = 0.3$ respectively. The *horizontal line* is the estimated asymptotic value (5.53)

Our estimates of the correlation length exponent and (universal) critical spanning number, obtained from the data at $p = 1/2$, are:

$$\nu = 2.745(19), \qquad n_s^{\text{crit}} = 2.035(10). \tag{5.53}$$

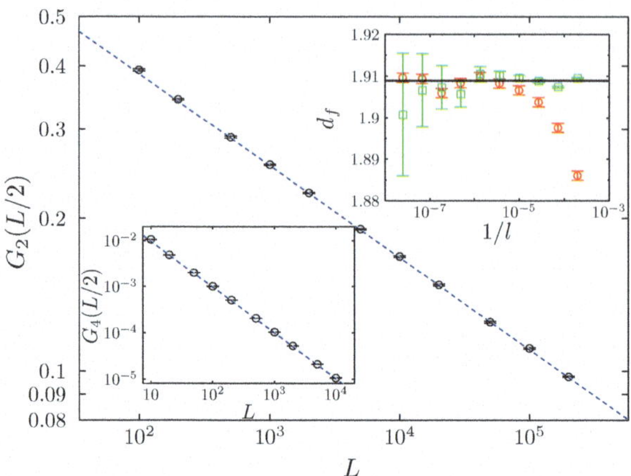

Fig. 5.15 *Main panel* the two-leg watermelon correlator G_2 at the critical point. The power-law decay $G_2 \sim L^{-2x_2}$ gives the fractal dimension d_f via (5.57). The *upper inset* compares this value (indicated by the *horizontal black line*) with the finite-size estimates coming from $\Delta X(l)$ (*red circles*) and $P(l)$ (*green squares*)—see text. The *lower inset* shows the four-leg watermelon correlator G_4

We cannot constrain the irrelevant exponent very precisely. From the full fit, we obtain

$$y_{\text{irr}} \in -(0.2, 0.35). \tag{5.54}$$

A direct estimate from the finite size corrections to the spanning number at the critical point gives a result compatible with this: the fit in the inset to Fig. 5.13 corresponds to $y_{\text{irr}} = -0.272$.

Results for $p = 0.3$ are consistent with our expectation that all points on the critical line are in the same universality class, but error bars are larger because of larger finite size effects and smaller system sizes. We find $\nu = 2.87(10)$ and $n_s^{\text{crit}} = 2.07(3)$. With regard to the convergence to a common n_s^{crit}, see the inset to Fig. 5.14 which shows the n_s-values of the crossings between curves for consecutive L values.

5.5.2 Watermelon Correlators, d_f, and Length Distribution

Next we consider the watermelon correlation functions G_2 and G_4 defined in Sect. 5.3.2. We evaluate these correlators at separation $L/2$ for a range of system sizes L—see Fig. 5.15. The data is for the critical point at $p = 1/2$.

Fitting to pure power laws,

$$G_k(L/2) \propto L^{-2x_k}, \tag{5.55}$$

we obtain the scaling dimensions of the two- and four-leg operators:

$$x_2 = 0.091(1), \quad x_4 = 0.491(1). \tag{5.56}$$

The scaling relation

$$d_f = 2 - x_2 \tag{5.57}$$

gives the fractal dimension of the critical loops,

$$d_f = 1.909(1). \tag{5.58}$$

We may obtain independent finite size estimates of d_f either using the numerical estimate for $\mathrm{d}\ln \Delta X/\mathrm{d}\ln l$ (where $\Delta X(l)$ is the mean linear size of a loop of length l) or using $\mathrm{d}\ln P(l)/\mathrm{d}\ln l$ and the scaling relation (5.49). These are shown in the inset to Fig. 5.15—both plots are consistent with (5.58).

5.5.3 RG Equations in the Presence of Vortices

For an approximate description of the transition, we extend the RG description (5.21) to take account of the nonzero fugacity for $\mathbb{Z}_2$ vortices. In this we follow the treatment by Fu and Kane of the $O(2N)/O(N) \times O(N)$ sigma model at $N \to 0$ [20]. This sigma model and the RP^{n-1} sigma model are similar—both sustain $\mathbb{Z}_2$ vortices, and each reduces to the XY model in an appropriate limit, which can be expanded around. An expansion around the XY limit was also considered for the $O(n)$ model near $n = 2$ by Cardy and Hamber [57]. The importance of topological defects in replica sigma models for localisation in two dimensions was also pointed out by Konig et al., who developed an RG approach to localisation in the chiral symmetry classes taking account of $\mathbb{Z}$ vortices [19].

Since $\mathrm{RP}^1 = S^1$, our sigma model coincides with the XY model at $n = 2$. With the normalisation of Eq. 5.19, this has a Kosterlitz Thouless transition at the critical stiffness $K_c = 8/\pi$. The RG flows near this point are governed by the Kosterlitz RG equations for K and the vortex fugacity, which we denote V.

We assume that we can expand the RG equations in $(2-n)$, and that V should be interpreted as the fugacity for $\mathbb{Z}_2$ vortices[9] when $n \neq 2$. At lowest order, these equations are corrected by the β-function for K in the absence of vortices, $\mathrm{d}K/\mathrm{d}\tau \simeq (2-n)f(K)$:

$$\frac{\mathrm{d}V}{\mathrm{d}\tau} = \left(2 - \frac{\pi K}{4}\right) V, \qquad \frac{\mathrm{d}K}{\mathrm{d}\tau} = (2-n) f(K_c) - V^2. \tag{5.59}$$

[9] For $n > 2$, a $\mathbb{Z}_2$ vortex corresponds to e.g. $\mathbf{S} = (\cos\theta/2, \sin\theta/2, 0, ..., 0)$; when $n = 2$ this configration becomes a $\mathbb{Z}$ vortex of charge ± 1. (Here θ is a polar coordinate; recall $\mathbf{S} \sim -\mathbf{S}$.)

These equations yield critical points at $K = K_c = 8/\pi$ and $V = \pm\sqrt{(2-n)f(K_c)}$, with critical exponents

$$\nu = \sqrt{\frac{2}{(2-n)\pi f(K_c)}}, \qquad y_{\text{irr}} = -\sqrt{\frac{(2-n)\pi f(K_c)}{2}}. \tag{5.60}$$

The critical stiffness $K_c \sim 2.4$ is of roughly the same magnitude as the critical winding number $n_s^{\text{crit}} \sim 2.0$ at $n = 1$ (5.53), as we expect from Sect. 5.4.2. Making the further approximation of evaluating Eq. 5.60 with the two-loop (large K) beta function, $f(K_c) \simeq (2\pi)^{-1}[1 + (2\pi K_c)^{-1}]$, we have at $n = 1$

$$\nu \sim 1.9, \qquad y_{\text{irr}} \sim -0.5. \tag{5.61}$$

As $n \to 2$, ν diverges and the irrelevant exponent tends to zero.

Unsurprisingly, this crude approximation does not give quantitatively accurate results for $n = 1$, but it does reproduce the qualitative structure of the phase diagram (with the Goldstone phase sandwiched between massive phases at positive and negative V) and the appearance of a large correlation length exponent.

A comparison with alternative approaches would be desirable. For example, is it possible to make an expansion in $(2 - n)$ while avoiding the additional large K approximation needed here?

The approximate RG treatment above can be extended to dimensions greater than two, in analogy to Ref. [57]. This may be a useful way to think about the 3D unoriented loop model (Sect. 2.10).

5.6 Outline of Numerical Methods

The results above are for system sizes from $L = 100$ up to $L = 10^6$. For sizes up to $L \sim 2 \times 10^4$ a straightforward Monte Carlo procedure was used, which of course benefits from the fact that node configurations are independent random variables when $n = 1$. Very large sizes require a more efficient 'knitting and shuffling' procedure.

The straightforward approach involves constructing independent $L \times L$ samples and calculating n_s, $P(l)$, $\Delta X(l)$, $G_2(L/2)$ and $G_4(L/2)$ by following the loops. Since at $n = 1$ the Boltzmann weight is independent of boundary conditions, the same samples yield both n_s (defined for a cylinder) and the other quantities (for which periodic BCs are used). G_2 is the probability that two links are connected, and G_4 is the probability that two nodes are visited by the same pair of loops. While this is not the only way for two nodes to be connected by four strands, which could make up a single loop, the scaling is the same for both types of contribution.

In the 'knitting' approach, larger samples are 'grown' by successive addition of $L \times L'$ strips, where L' is of order one. The independence of the nodes at $n = 1$ means that it is not necessary to store the full configuration, only the connectivity of the boundary links and the lengths of the connecting strands. Thus the required memory is $O(L \ln L)$ rather than $O(L^2)$. When the addition of a strip closes a loop, its length and height in the growing direction (ΔX) are added to histograms. (Cylinders with circumference 10^6 and height much larger than this were considered.)

This method requires storing the connectivity information for the boundary links of a large cylinder. This has the flavour of a transmission matrix in a localisation problem—in the future it would be interesting to consider the properties of this 'matrix' in more detail. Note that the connectivity information is configuration-dependent—this is unlike the transfer matrix approach, in which the transfer matrix is not configuration-dependent, and which is limited to small sizes.

Finally, we can improve efficiency by 'shuffling'. We knit many strips of width L and height $H = L/20$. For each strip we store the connectivity of the boundary links. Joining 20 of them yields a square sample, and we have enough information to calculate G_2 (for links lying on the boundaries of strips) and n_s. From each set of 20 pieces, many different samples may be created by shuffling the order of the pieces and by rotating them in the transverse direction. 1,000 samples are constructed for each set. These are not of course independent, so error bars are estimated by producing many independent such sets (80–200) and examining the statistical fluctuations between them. For more details see Ref. [13].

5.7 Outlook

The loop models we have discussed are described by 'replica' sigma models of the kind familiar from localization and polymer physics. Such problems remain at the frontier of our understanding of critical phenomena, and we hope that the transitions in the loop models will provide a testing ground for new approaches.

While exact results for the critical behaviour discussed in Sect. 5.5 would be desirable, the development of more accurate analytical approximations would also be enlightening. On the numerical side, work on the critical loop model should be extended to other values of n, either via Monte Carlo or the transfer matrix [23], in order to pin down the properties of the whole family of critical points for $0 < n < 2$. Note also that while taking the limit $n \to 0$ for fixed p, q leaves us in the Goldstone phase, sending $n, p, q \to 0$ at the same time could leave an interesting critical model at $n = 0$, in which a subset of the loop configurations is retained. We hope to return to these issues.

The connection between the CPLC at $n = 1$ and disordered fermions remains an open question. To begin with, recall the situation for the loop model without crossings ($p = 0$). This can be related to localisation in at least two ways. Firstly, a limiting case of the Chalker Coddington model for the quantum Hall effect [58], in which the scattering matrices at a node become 'classical', yields the loop model without crossings—i.e. classical percolation. This is the familiar semiclassical description of the quantum Hall transition [59], but because quantum tunnelling has not been taken into account, it does not correctly capture the universal critical behaviour. However, the loop model has a second relationship with localisation which is less obvious and which does not rely on suppressing quantum tunnelling. This is due to an exact mapping from a network model for the spin quantum Hall transition [40] (an analogue of the quantum Hall transition, but in symmetry class C rather than A [39]) to the loop model [25, 26, 30, 36].

For loops with crossings ($p > 0$) we can again construct a mapping of the first kind by taking a 'classical' limit in a network model with a Kramers doublet on each edge (so that, in a given realisation of disorder, each quantum node corresponds to one of the three classical possibilities in Fig. 5.2). While this correspondence captures a surprising number of features of the phase diagram for localisation in the symplectic symmetry class (e.g. logarithmic flow of the conductance in the metallic phase), it is a somewhat trivial one because of the explicit suppression of quantum tunnelling between different loops. It would be interesting to know whether the analogy with localisation goes beyond this—in particular, whether the critical behaviour of the loop model can be related to the critical behaviour of a true localisation problem. In this regard it is interesting to note that our value of ν is very close to estimates of ν for the symplectic class [60–63].

Returning to loop models in their own right, there is a good understanding of the zoology of critical points in models without crossings, many of which fit into the one-parameter family of universality classes in SLE_κ. In general, crossings take us outside this family. Here we have discussed a new line of critical points exemplifying this, but we certainly do not expect that this exhausts the possibilities for new critical behaviour—much remains to be learned.

References

1. P.G. de Gennes, Phys. Lett. A **38**, 339 (1972)
2. A.M.M. Pruisken, Nucl. Phys. B **235**, 277 (1984)
3. H. Levine, S.B. Libby, A.M.M. Pruisken, Phys. Rev. Lett. **51**, 1915 (1983)
4. A.J. McKane, Phys. Lett. A **76**, 22 (1980)
5. G. Parisi, N. Sourlas, J. Phys. Lett. **41**, L403 (1980)
6. H. Weidenüller, Nucl. Phys. B **290**, 87 (1987)
7. F. Evers, A.D. Mirlin, Rev. Mod. Phys. **80**, 1355 (2008)
8. P. Fendley, in *New Theoretical Approaches to Strongly Correlated Systems*, ed. by A.M. Tsvelik (Kluwer, Dordrecht, 2001) [cond-mat/0006360]
9. C. Candu, J.L. Jacobsen, N. Read, H. Saleur, J. Phys. A **43**, 142001 (2010)
10. N. Read, H. Saleur, Nucl. Phys. B **613**, 409 (2001)
11. J.L. Jacobsen, N. Read, H. Saleur, Phys. Rev. Lett. **90**, 090601 (2003)
12. A. Nahum, J.T. Chalker, P. Serna, M. Ortuño, A.M. Somoza, Phys. Rev. Lett. **107**, 110601 (2011)
13. A. Nahum, P. Serna, A.M. Somoza, M. Ortuño, Phys. Rev. B **87**, 184204 (2013)
14. B. Nienhuis, in *Phase Transitions and Critical Phenomena*, Chapter 1, vol 11, ed. by C. Domb, J. Liebowitz (Academic Press, Singapore, 1987)
15. J. Cardy, Ann. Phys. **318**, 81 (2005)
16. A.L. Owczarek, T. Prellberg, J. Stat. Phys. **79**, 951 (1995)

17. D.P. Foster, J. Phys. A Math. Theor. **42**, 372002 (2009)
18. R.M. Ziff, X.P. Kong, E.G.D. Cohen, Phys. Rev. A **44**, 2410 (1991)
19. E.J. Konig, P.M. Ostrovsky, I.V. Protopopov, A.D. Mirlin, Phys. Rev. B **85**, 195130 (2012)
20. L. Fu, C.L. Kane, Phys. Rev. Lett. **109**, 246605 (2012)
21. M.J. Martins, B. Nienhuis, R. Rietman, Phys. Rev. Lett. **81**, 504 (1998)
22. W. Kager, B. Nienhuis, J. Stat. Mech. P08004 (2006)
23. Y. Ikhlef, J. Jacobsen, H. Saleur, J. Stat. Mech. P05005 (2007)
24. J.W. Lyklema, J. Phys. A Math. Gen. **18**, L617 (1985)
25. I.A. Gruzberg, A.W.W. Ludwig, N. Read, Phys. Rev. Lett. **82**, 4524 (1999)
26. E.J. Beamond, J. Cardy, J.T. Chalker, Phys. Rev. B **65**, 214301 (2002)
27. M. Ortuño, A.M. Somoza, J.T. Chalker, Phys. Rev. Lett. **102**, 070603 (2009)
28. Y. Ikhlef, P. Fendley, J. Cardy, Phys. Rev. B **84**, 144201 (2011)
29. E. Bettelheim, I.A. Gruzberg, A.W.W. Ludwig, Phys. Rev. B **86**, 165324 (2012)
30. J. Cardy, in *50 Years of Anderson Localization*, ed. by E. Abrahams (World Scientific, Singapore, 2010)
31. D.J. Thouless, J. Phys. C **8**, 1803 (1975)
32. V. Gurarie, Nucl. Phys. B **410**, 535 (1993)
33. A. Gainutdinov, J.L. Jacobsen, N. Read, H. Saleur, R. Vasseur, J. Phys. A Math. Theor. **46**, 494012 (2013)
34. J. Cardy, J. Phys. A Math. Theor. **46**, 494001 (2013)
35. V. Gurarie, J. Phys. A Math. Theor. **46**, 494003 (2013)
36. A.D. Mirlin, F. Evers, A. Mildenberger, J. Phys. A Math. Gen. **36**, 3255 (2003)
37. T. Senthil, M.P.A. Fisher, L. Balents, C. Nayak, Phys. Rev. Lett. **81**, 4704 (1998)
38. R. Bundschuh, C. Cassanello, D. Serban, M.R. Zirnbauer, Phys. Rev. B **59**, 4382 (1999)
39. T. Senthil, J.B. Marston, M.P.A. Fisher, Phys. Rev. B **60**, 4245 (1999)
40. V. Kagalovsky, B. Horovitz, Y. Avishai, J.T. Chalker, Phys. Rev. Lett. **82**, 3516 (1999)
41. E.J. Beamond, A.L. Owczarek, J. Cardy, J. Phys. A Math. Gen. **36**, 10251 (2003)
42. J.M.F. Gunn, M. Ortuño, J. Phys. A Math. Gen. **18**, L1095 (1985)
43. K. Shtengel, L.P. Chayes, J. Stat. Mech. P07006 (2005)
44. J.T. Chalker, M. Ortuño, A.M. Somoza, Phys. Rev. B **83**, 115317 (2011)
45. N. Read, H. Saleur, Nucl. Phys. B **777**, 263 (2007)
46. N.D. Mermin, Rev. Mod. Phys. **51**, 591 (1979)
47. P.E. Lammert, D.S. Rokhsar, J. Toner, Phys. Rev. Lett. **70**, 1650 (1993)
48. A.M.P. Silva, A.M.J. Schakel, G. L. Vasconcelos, (2013), arXiv:1306.4996
49. A.M. Polyakov, Phys. Lett. B **59**, 79 (1975)
50. I. Affleck, Nucl. Phys. B **257**, 397 (1985)
51. S. Hikami, Phys. Lett. **98B**, 208 (1981)
52. I. Affleck, Phys. Rev. Lett. **56**, 408 (1986)
53. T. Senthil, M.P.A. Fisher, Phys. Rev. B **74**, 064405 (2006)
54. R.A. Pelcovits, D.R. Nelson, Phys. Lett. **57 A**, 23 (1976)
55. D.R. Nelson, R.A. Pelcovits, Phys. Rev. B **16**, 2191 (1977)
56. J. Cardy, in *Fluctuating Geometries in Statistical Mechanics and Field Theory*, ed. by F. David, P. Ginsparg, J. Zinn-Justin, (Elsevier, Amsterdam, 1996) [cond-mat/9409094]
57. J.L. Cardy, H.W. Hamber, Phys. Rev. Lett. **45**, 499 (1980)
58. J.T. Chalker, P.D. Coddington, J. Phys. C. **21**, 2665 (1988)
59. S.A. Trugman, Phys. Rev. B **27**, 7539 (1983)
60. Y. Asada, K. Slevin, T. Ohtsuki, Phys. Rev. Lett. **89**, 256601 (2002)
61. P. Markos, L. Schweitzer, J. Phys. A **39**, 3221 (2006)
62. A. Mildenberger, F. Evers, Phys. Rev. B **75**, 041303 (2007)
63. H. Obuse, A.R. Subramaniam, A. Furusaki, I.A. Gruzberg, A.W.W. Ludwig, Phys. Rev. B **82**, 035309 (2010)

Chapter 6
Polymer Collapse

6.1 Introduction

A long polymer with repulsive or excluded volume interactions between segments displays the universal behaviour of the lattice self-avoiding walk (SAW), while strong enough attractive interactions cause the polymer to collapse. The boundary between these two regimes is the so-called Θ point. In de Gennes' description of the polymer via the $O(N)$ model at $N = 0$, the SAW corresponds to the critical point, and the Θ point to the tricritical point [1–3]. However the full picture is more complicated, especially in the two dimensional case, which we now discuss.

While the SAW behaviour is extremely robust, the Θ point is more subtle. Different lattice models can show collapse transitions in different universality classes [4–7], and for some models the structure of the phase diagram near the Θ point is more complex than would be expected from the de Gennes picture [4, 7–9]. Additionally, despite many years of theoretical debate and numerical simulations, a framework for classifying the many models that have been studied has been lacking. As a result the issue of which models capture the generic critical behaviour of the Θ polymer, and which should be regarded as fine-tuned, has not been fully understood. Numerical results [10–13] suggest that the exact exponents derived by Duplantier and Saleur for a particular model without crossings [5, 14] are generic for models without crossings, but even this issue is not yet understood.

The best-studied model for a polymer that *can* cross itself is the 'interacting self-avoiding trail', or ISAT. This model has been particularly controversial [6, 7, 15–18]. Numerous conflicting results and hypotheses have been put forward for the critical exponents at the collapse point, and the phase diagram has not been understood.

Here we will construct a field theory for the ISAT, making use of the results of Chap. 5, and use it to resolve the confusion regarding exponents and explain the

A. Nahum, *Critical Phenomena in Loop Models*, Springer Theses,
DOI 10.1007/978-3-319-06407-9_6

numerical phase diagram [7]. This treatment also shows that the ISAT Θ point[1]—despite the simplicity and naturalness of the model—is highly fine-tuned from the point of view of more general polymer models. Specifically, it is an infinite-order multicritical point at which the symmetry of the problem is enhanced from the $O(N \to 0)$ symmetry of a generic model to $O(N \to 1)$. The presence of additional symmetry is inseparable from some of the desirable features of the model, in particular its amenability to very high precision numerical work, and is a general feature of models that are related to 'smart walks' (in any number of dimensions).

Having understood the space of polymer models in the vicinity of the ISAT, we briefly speculate on the classification of more general models for polymer collapse. Our analysis indicates that the Θ point exponents of a 'generic' model with crossings are still unknown. Models which should show the generic behaviour have been studied, but currently the numerical precision is not sufficient to say whether, for example, the fractal dimension at the Θ point differs from the Duplantier Saleur value [19, 20]. Note that the inclusion of crossings in a 2D model is quite natural; for example, they will be present if the model describes a polymer confined to a quasi-2D geometry.

6.2 The Interacting Self-Avoiding Trail

The ISAT polymer can visit links of the square lattice only once, but nodes twice, and the interaction strength is determined by the weight associated with each doubly-visited node (that is, each self-contact). For simplicity we consider ring polymers, rather than open ones; allowed polymer configurations are then equivalent to allowed configurations of a *single* loop in the CPLC. We place the polymer on a lattice of large finite linear size L in order to make the partition sum finite. It is:

$$Z_{\text{pol}}(k,t) = \sum_{\substack{\text{polymer}\\ \text{configs}}} k^{(\text{length})} t^{(\text{no. self-contacts})}. \tag{6.1}$$

Translated versions of a loop are counted as distinct polymer configurations in this sum. Note that we are now in the ensemble with fixed length fugacity k, rather than fixed length; this is more convenient from the point of view of field theory, but the two ensembles are simply related [3]. The parameter t controls interactions which are repulsive for $t < 1$ and attractive for $t > 1$.

An interesting feature of this model is its phase diagram, obtained by Foster from numerical transfer matrix calculations [7]. A schematic version is shown in Fig. 6.1. For small k the polymer is of finite typical size and not critical—the 'zero density' phase. For large k, the polymer has a length of order L^2 and visits a finite fraction

[1] We use the term 'Θ point' synonymously with 'collapse point', rather than to refer to a single RG fixed point. Thus different models may show 'Θ points' in different universality classes, which we will refer to as the ISAT Θ point, the Duplantier Saleur Θ point, et cetera.

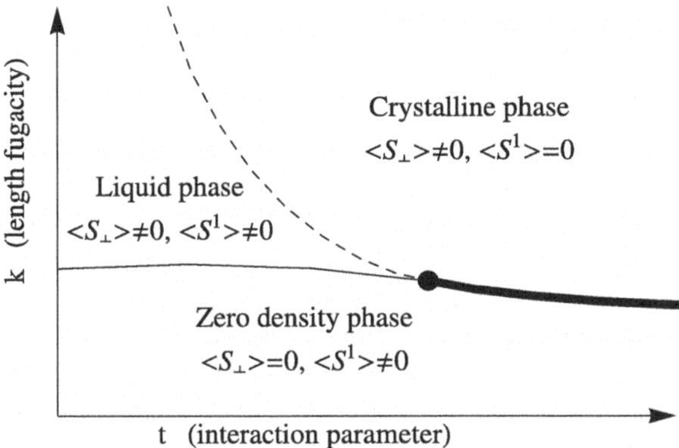

Fig. 6.1 Schematic representation of the phase diagram for the polymer found in Ref. [7], together with our interpretation in terms of a perturbed $O(1+n')$ model in the limit $n' \to 0$. The *thin solid line* is in the universality class of the self-avoiding walk (polymer in good solvent), or critical $O(n')$ model. The *dot* is the multicritical collapse point, or Θ point, with full $O(1+n')$ symmetry. The *bold line* is a first-order transition. The *dashed line* is a transition in the Ising universality class between two regimes in which the polymer is dense (visits a finite fraction of the links of the lattice)—the 'crystalline' and 'liquid' phases

of the links of the lattice: this is the dense regime, which includes a 'liquid' and a 'crystalline' phase.

The behaviour of a polymer of fixed large length on an infinite lattice is governed by the singularity of the partition function closest to $k = 0$, i.e. by a point on the transition line separating the zero-density and dense regimes. When t is small, this transition shows the usual critical behaviour of the SAW (the thin solid line in Fig. 6.1). For large t there is a first order transition between zero density and dense phases (thick line), which is associated with the collapsed polymer.[2] The Θ point (dot) separates these.

An unexpected feature, from the point of view of the de Gennes theory, is an additional line of transitions within the dense regime (dashed line) which are found numerically to be in the Ising universality class [7]. This Ising transition separates the crystalline and liquid phases, which we will characterise later on; the polymer is critical in both of these phases.

A similar line of Ising transitions terminating at a Θ point was found in a polymer model without crossings studied by Blöte and Nienhuis [4], and a heuristic explanation for it was provided by associating Ising degrees of freedom (for which the polymer was a domain wall) with the faces of the lattice [4, 21]. In that model, the absence of crossings also allows for a Coulomb gas description which captures the Ising transition [22].

---

[2] That is, the collapsed polymer should be regarded as a domain of the dense phase within a background of the zero-density phase. The volume of this domain is determined by the length of the polymer.

6.3 Field Theory for the ISAT

We now make use of the fact that the ISAT, when precisely at its Θ point, can be mapped to the CPLC at a point in its Goldstone phase. (The ISAT Θ point is at $k = 1/3$, $t = 3$ [6], and the corresponding point for the CPLC is $n = 1$, $p = 1/3$, $q = 1/2$.) The Θ point is therefore described by the $O(n \to 1)$ sigma model, and we can understand the region around it by perturbing this sigma model. The required perturbations can be written down explicitly in the lattice model of Sect. 5.3.5, and the continuum versions follow by straightforward coarse-graining. The considerations below may readily be generalised to the various modifications of ISAT that have been studied numerically, e.g. on other lattices [8], with additional interactions [19, 20, 23, 24], or in higher dimensions [25, 26].

Before continuing, we note that an alternative conjecture for the critical behaviour of the ISAT Θ point was put forward on the basis of numerical transfer matrix calculations in Ref. [7]. According to this conjecture, the Θ point has nontrivial critical exponents identical to those of an exactly solvable model of polymers *without* intersections [4, 27]. For example, a thermal exponent $\nu = 12/23$ was conjectured for the ISAT Θ-point [7]. This is at odds with the predictions of the Goldstone phase, which yields trivial critical exponents together with universal logarithmic corrections. We believe the Goldstone phase scenario is convincingly established by our numerical results for large systems, together with the logarithmic behaviour seen numerically in Refs. [6, 28] and the theoretical arguments here and in Ref. [29], and that the apparent nontrivial exponents in the transfer matrix calculation of Ref. [7] are due to logarithmic finite size corrections. The form of these corrections may be extracted from the sigma model.[3]

To relate the single-polymer problem to the spin model, we proceed along similar lines to Sect. 5.4.3, writing $\mathbf{S} = (S^1, \mathbf{S}_\perp)$ and expanding the partition function for the CPLC in $n' = n - 1$. This separates out configurations with a single marked loop, or in other words a single $\mathbf{S}_\perp$ worldline, and this will be our polymer (see Fig. 6.2.)

[3] In more detail, we use the sigma model to estimate the eigenvalues of the transfer matrix for a cylinder of circumference L. Coarse-graining by a factor L yields an effective 1D sigma model with renormalised stiffness K_L. This is equivalent to the quantum mechanics of a particle on an $(n-1)$-dimensional sphere. In the limit $n \to 1$, the energy levels are

$$E_k = \frac{k(k-1)}{2K_L}, \quad k = 0, 1, 2, \ldots \tag{6.2}$$

This leads to the finite size estimates $x_k = E_k/2\pi$ for the scaling dimensions, which tend to zero logarithmically with L:

$$x_k \sim \frac{k(k-1)}{2 \ln L/L_0}. \tag{6.3}$$

However this asymptotic formula is not very relevant to numerical transfer matrix calculations, which are restricted to small sizes. For a crude estimate of the expected magnitude of the finite size exponents, we can instead approximate K_L by the spanning number on an $L \times L$ cylinder, which we have calculated numerically at the relevant point $p = 1/3$, $q = 1/2$. For L from 5 to 12, this gives x_2 varying from 0.103 to 0.086.

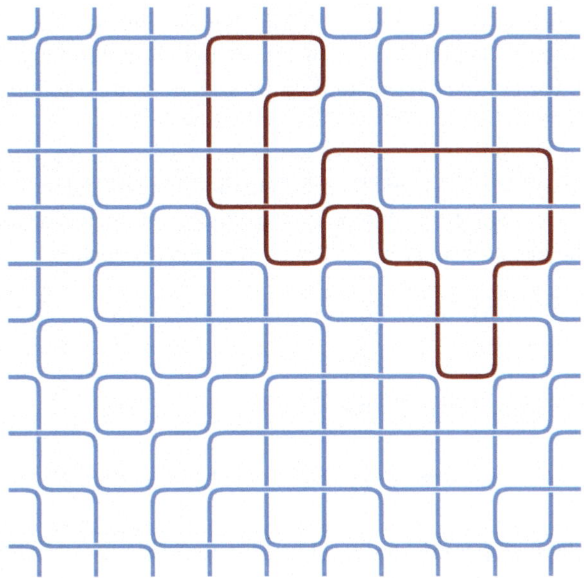

Fig. 6.2 The ISAT polymer at its Θ point behaves like a marked (*dark red*) loop in the CPLC at $n = 1$, $p = 1/3$, $q = 1/2$. *Dark red* loops are associated with worldlines of $\mathbf{S}_\perp$, and *light blue* loops with worldlines of S^1

In order to control k and t for this polymer we must modify the partition function (5.22) in a way that breaks the symmetry of the $O(n)$ sigma model down to $\mathbb{Z}_2 \times O(n')$. At each node, (5.22) contains terms of the form $(\mathbf{S}_1.\mathbf{S}_2)(\mathbf{S}_3.\mathbf{S}_4)$. Such a term corresponds to two sections of loop passing through the node, one connecting link 1 to link 2, and one connecting link 3 to link 4. In the new ensemble, these sections can be sections of marked or unmarked loop, and we modify the weights accordingly:

$$\begin{aligned}(\mathbf{S}_1.\mathbf{S}_2)(\mathbf{S}_3.\mathbf{S}_4) \longrightarrow \; & (S_1^1 S_2^1)(S_3^1.S_4^1) \\ & + 3k\,(S_1^1.S_2^1)(\mathbf{S}_{\perp 3}.\mathbf{S}_{\perp 4}) + 3k\,(\mathbf{S}_{\perp 1}.\mathbf{S}_{\perp 2})(S_3^1.S_4^1) \\ & + 3k^2 t\,(\mathbf{S}_{\perp 1}.\mathbf{S}_{\perp 2})(\mathbf{S}_{\perp 3}.\mathbf{S}_{\perp 4}). \end{aligned} \tag{6.4}$$

Each unit of marked length then acquires a factor of $3k$, and each meeting of two marked strands acquires an additional factor of $t/3$. Expanding the partition function in n' as in Eqs. (5.39) and (5.41) gives

$$Z(n', k, t) = 1 + n' \sum_{\substack{\mathcal{C};\text{ one} \\ \text{marked loop}}} W_{\mathcal{C}}\,(3k)^{\text{length}}(t/3)^{\text{no. self-contacts}} + \dots \tag{6.5}$$

Separating the sum into a sum over configurations of the marked loop (polymer) and a sum over the configurations of the other loops, and performing the latter, gives

$$Z(n', k, t) = 1 + n'\, Z_{\text{pol}}(k, t) + \dots, \qquad Z_{\text{pol}}(k, t) = \partial_{n'} Z(n', k, t)\big|_{n'=0} \tag{6.6}$$

which is the required relation. For brevity we have discussed only the partition function, but one may easily check that the natural geometrical correlation functions in the polymer problem, i.e. the watermelon correlators for the polymer, are given by correlators of $\mathbf{S}_\perp$ in the replica limit $n' \to 0$.

The polymer multicritical point at $(k, t) = (\frac{1}{3}, 3)$ corresponds to the CPLC at $(p, q) = (\frac{1}{3}, \frac{1}{2})$, and therefore to the sigma model with full $O(n)$ symmetry. Varying (k, t) away from this point introduces symmetry-breaking perturbations, of which the most relevant are the two lowest anisotropies:

$$O_{\perp 1} = \mathbf{S}_\perp^2, \quad O_{\perp 2} = (\mathbf{S}_\perp^2)^2 - \frac{2(n+1)}{n+4}\mathbf{S}_\perp^2. \tag{6.7}$$

The perturbed Lagrangian for the sigma model is

$$\mathcal{L} = \frac{K}{2}\Big((\nabla \mathbf{S})^2 + \gamma_1 O_{\perp 1} + \gamma_2 O_{\perp 2}\Big) \qquad (\mathbf{S}^2 = 1). \tag{6.8}$$

The bare parameters γ_1 and γ_2 are linearly related to the perturbations

$$\delta k = k - 1/3, \quad \delta t = t - 3 \tag{6.9}$$

when these are small. We can obtain a very crude estimate of the relation from the explicit lattice partition function $Z(n', k, t)$. We evaluate the right hand side of Eq. (6.4) for spatially constant $\mathbf{S}_l$, and take the logarithm of the Boltzmann weight for a node to obtain the potential terms in the bare Lagrangian. This gives

$$\gamma_1 \sim -C\left(\delta k + \frac{2}{45}\delta t\right), \quad \gamma_2 \sim -C\frac{\delta t}{18}, \tag{6.10}$$

where C is an undetermined positive constant.

We now use (6.8) to explain the phase diagram of the ISAT, Fig. 6.1.

6.4 Phase Diagram for the ISAT

Before considering the more detailed RG picture, we identify the phases in Fig. 6.1 with the phases of this perturbed sigma model. These are characterized by whether the $\mathbb{Z}_2$ and $O(n')$ symmetries are broken, i.e. whether $\langle S^1 \rangle \neq 0$ and whether $\langle \mathbf{S}_\perp \rangle \neq 0$. Since the Θ point is controlled by the fixed point at infinite stiffness, the phase with $\langle S^1 \rangle = \langle \mathbf{S}_\perp \rangle = 0$ does not appear upon perturbing around it. However the other three do.

The leading effect of reducing the length fugacity k below the Θ-point value is to introduce a mass for $\mathbf{S}_\perp$, so that $\mathbf{S}$ orders in the longitudinal direction and the transverse modes $\mathbf{S}_\perp$ are massive. This gives the *zero-density phase*,

$$\langle \mathbf{S}_\perp \rangle = 0, \quad \langle S^1 \rangle \neq 0. \tag{6.11}$$

Correlators decay exponentially for the polymer, and it has a finite typical size.[4]

Moving away from the Θ-point by increasing the length fugacity makes S^1 massive, and **S** 'orders' in the transverse plane:

$$\langle \mathbf{S}_\perp \rangle \neq 0, \quad \langle S^1 \rangle = 0. \tag{6.12}$$

This is the *dense phase with Ising disorder*, or the crystalline phase. The polymer fills the system densely, and the transverse modes $\mathbf{S}_\perp$ are in a Goldstone phase. To see that this is possible, note that when fluctuations in S^1 are massive we may imagine integrating them out to get an effective $O(n')$ sigma model for $\mathbf{S}_\perp$. In the limit $n' \to 0$, this sigma model has a Goldstone phase as a consequence of the beta function in Eq. (5.21). This is just the theory for the dense polymer with crossings studied in Ref. [29]. The two-leg watermelon correlation function is a constant at long distances, as discussed in Sect. 5.4.1, meaning that the polymer is dense.

For appropriate values of the coefficients γ_1, γ_2, we will find that the renormalised free energy is minimised by taking both $\langle \mathbf{S}_\perp \rangle$ and $\langle S^1 \rangle$ nonzero, so that both symmetries, $\mathbb{Z}_2$ and $O(n')$, are broken:

$$\langle \mathbf{S}_\perp \rangle \neq 0, \quad \langle S^1 \rangle \neq 0. \tag{6.13}$$

This is the *dense phase with Ising order*, or the liquid phase. Again the transverse modes are in the Goldstone phase of the $O(n' \to 0)$ model. We will return to what Ising order/disorder means for the polymer below.[5]

The field theory also determines the nature of the phase transitions. First, consider the thin solid line in Fig. 6.1. The field S^1 is Ising-ordered on both sides of this transition, and its massive fluctuations play no role in the critical behaviour of $\mathbf{S}_\perp$. We therefore have the critical point of the $O(n')$ model in the limit $n' \to 0$. This is the usual description of the self-avoiding walk [1, 3], confirming what we expect and indeed find in the polymer problem.

Next, consider the dashed line in Fig. 6.1. Here, S^1 undergoes an ordering transition at which $\mathbb{Z}_2$ symmetry is broken. Thus we would expect an Ising transition, but

---

[4] In the zero density phase the *un*marked loops remain in the Goldstone phase, being unaffected by the polymer since it occupies an infinitesimal fraction of the system volume (in the thermodynamic limit). In order to write down watermelon correlation functions for the unmarked loops, we would have to extend S^1 to a vector $\mathbf{S}_1$ with $m \to 1$ components, increasing the symmetries of the field theory to $O(m) \times O(n')$.

[5] A minor subtlety regarding the above classification is that only gauge-invariant operators are meaningful in the loop model/RP^{n-1} model. For this reason the global symmetry of the perturbed loop model is $O(n') = \mathbb{Z}_2 \times SO(n')$, rather than $\mathbb{Z}_2 \times O(n')$ as in the perturbed $O(n)$ sigma model. (The zero-density phase breaks no symmetries of the perturbed RP^{n-1} model, the 'dense phase with Ising order' fully breaks $O(n')$, and the 'dense phase with Ising disorder' breaks only $SO(n')$.) However we are free to use the language of the latter, which (for the reasons discussed in Sect. 5.3.4) captures the universal properties of the perturbed RP^{n-1} sigma model in the regime we are considering.

we must check that the massless degrees of freedom associated with $\mathbf{S}_\perp$—which are in the Goldstone phase—do not modify the Ising critical behaviour. It is easy to see they do not. The most relevant coupling allowed by $\mathbb{Z}_2 \times O(n')$ symmetry is via the product of the energy operators,

$$E_{\text{Ising}} \times E_{\text{Goldstone}}. \tag{6.14}$$

This composite operator has dimension (length)$^{-3}$, so is irrelevant with $y_{\text{irr}} = -1$.

Finally, consider the thick line in the figure, which separates phases breaking different symmetries—$\mathbb{Z}_2$ on one side, and $O(n')$ on the other. According to Landau theory this transition should be first order. This is confirmed in the more explicit approach below and agrees with the numerical findings [7].

What does Ising order/disorder mean for the polymer? Let us define a new configuration of Ising spins μ_F on the faces F of the square lattice by the requirement that the polymer is the (only!) domain wall in this configuration. Then the dense phase with Ising *disorder* corresponds to antiferromagnetic *order* in μ, while μ is disordered in the dense phase with Ising order. (To show this, we write $\langle \mu_F \mu_{F'} \rangle$ in terms of a correlator of twist fields in the RP^{n-1} model, which force the fields Q^{1a} with $a > 1$ to change sign on a line connecting F and F'.) Antiferromagnetic order in μ is equivalent to the 'crystalline' order of Ref. [7], and it becomes perfect when the polymer visits every link of the lattice. It is also essentially equivalent to the Ising order defined in Refs. [4, 21] for a different model.

6.5 More Detailed Picture of Phase Diagram

For a more detailed picture of the phase diagram, we use the RG equations for the sigma model with lowest two anisotropies (Eq. (6.8)). After running the RG up to a large time τ_*, we find

$$K_* \sim \frac{\tau_*}{2\pi}, \quad \gamma_{1*} \sim \frac{\gamma_1 e^{2\tau_*}}{(\tau_*/2\pi K)^2}, \quad \gamma_{2*} \sim \frac{\gamma_2 e^{2\tau_*}}{(\tau_*/2\pi K)^7}.$$

For generic small initial values (γ_1, γ_2), we will renormalise to a regime where $\gamma_{1*} = O(1)$ and $|\gamma_{2*}| \ll |\gamma_{1*}|$, putting us deep within one of the phases—either the zero density phase or the dense phase with Ising disorder, depending on the sign of γ_1. The phase transitions occur instead in the regime where the renormalised γ_{1*} and γ_{2*} become of order one simultaneously.

Since the stiffness of the renormalised sigma model is large ($\tau_* \sim \ln |\gamma_1|^{-1/2} \sim \ln |\gamma_2|^{-1/2}$) we may determine which phase it is in simply by minimising the potential in the renormalised Lagrangian. In doing this we must bear in mind the constraint $\mathbf{S}^2 = 1$. We find the three phases described above, with the phase transition lines located at:

$$\text{SAW}: \quad \gamma_1 \simeq \frac{4}{5}\gamma_2 \left(\frac{4\pi K}{\ln 1/|\gamma_2|}\right)^5 \quad (\gamma_2 > 0),$$

$$\text{Ising}: \quad \gamma_1 \simeq -\frac{3}{5}\gamma_2 \left(\frac{4\pi K}{\ln 1/|\gamma_2|}\right)^5 \quad (\gamma_2 > 0),$$

$$1^{\text{st}}\text{order}: \quad \gamma_1 \simeq -\frac{1}{5}\gamma_2 \left(\frac{4\pi K}{\ln 1/|\gamma_2|}\right)^5 \quad (\gamma_2 < 0).$$

These formulas are valid asymptotically close to the Θ point. The above confirms that the three phases meet at the Θ-point, and shows that the SAW and Ising critical lines are asymptotically parallel as they approach the Θ point.

6.6 Generic Perturbations to the ISAT

An important question about the ISAT is whether it captures the generic Θ point behaviour for a polymer with crossings, i.e. whether the critical behaviour is robust to modifications of the model. These perturbations may break $O(n)$ symmetry but must preserve $O(n')$ symmetry. (Recall that $n' = n - 1$.)

The generic Θ point should have *two* relevant perturbations: although in an idealised experiment only the strength of the interactions must be tuned in order to reach the Θ point, another parameter is tuned to criticality automatically by taking the length of the polymer to infinity (see Sect. 6.2).

Therefore if the operators $O_{\perp 1}$ and $O_{\perp 2}$ of (6.7) were the only relevant perturbations of the ISAT Θ point, it would be a generic description of the collapse transition. But generic perturbations of the ISAT Θ point will introduce the entire infinite tower of higher anisotropies, $O_{\perp k} = (\mathbf{S}_\perp^2)^k + \cdots$, with couplings γ_k. At $n = 1$, the RG equation for γ_k is (to lowest order)

$$\frac{\mathrm{d}\gamma_k}{\mathrm{d}\tau} = \left(2 - \frac{2k^2 - k + 1}{2\pi K}\right)\gamma_k, \tag{6.15}$$

so all these perturbations are relevant at the $K = \infty$ fixed point. From the point of view of the polymer, this is therefore an infinite-order multicritical point and so cannot capture the generic universal behaviour of the Θ point polymer with crossings. This explains why the model analysed numerically in Ref. [30] shows different behaviour to the ISAT; a similar argument explains why the three-dimensional ISAT [25, 26] shows distinct universal behaviour from standard models of polymer collapse (Sect. 6.8).

The above indicates that if we extend the parameter space of the ISAT with an extra interaction, the ISAT Θ point will mark the point at which the collapse transition changes from being in the universality class of the tricritical $O(n' \to 0)$ model to being first order (at a cartoon level, it is the point where the sign of the $(\mathbf{S}_\perp^2)^3$

term changes in an appropriate renormalised Lagrangian). Recent numerical work by Bedini et al., who supplement the ISAT with an additional nearest neighbour interaction, confirms the appearance of a first order transition for one sign of the extra interaction and of a continuous one with non-ISAT exponents for the other sign [31].

6.7 Smart Walks and More General Models

Whether or not a model is fine-tuned of course depends on what it is supposed to represent. As a loop model with fugacity one, the CPLC is not fine-tuned. But the $O(n \to 1)$ symmetry that is guaranteed in such a problem is not guaranteed for the polymer, which generically has only an $O(n' \to 0)$ symmetry. This is why there is no contradiction in the fact that the ISAT Θ point, which is a multicritical point, maps to a point in the middle of a phase in the CPLC.

It is not a coincidence that many of the most studied models for Θ point polymers, including the ISAT and a number of others, permit mappings to models of classical disordered systems. In these mappings, the polymer, assuming it is a loop, becomes one of many loops in a configuration of a random system with short range correlations (like the short-range correlations between nodes in the CPLC at $n = 1$). As for the ISAT, these mappings hold only for particular values of the polymer interactions.

Such models are sometimes referred to as 'smart walks'. The correspondence with an uncorrelated random system means that polymer configurations can be 'grown' with the correct probabilities by sequentially fixing the state of the degrees of freedom at the growing tip. For example, starting with a given link in the CPLC, the smart walk makes a random decision about how (or whether) to turn every time it encounters a node that it has not already visited. This 'growability' property is very attractive for Monte Carlo, since long polymers can be rapidly generated. By contrast, for a polymer model with a generic Boltzmann weight, growing a loop by random *local* decisions at the tip produces configurations with probabilities that differ from the desired ones by a factor that grows exponentially with the length. (Ways around this obstacle are reviewed in Ref. [32].)

The best-known and historically most important 'smart' model is related to 2D percolation, and yields the exact Duplantier Saleur exponents for non-crossing Θ point polymers [5, 14]. In this model, defined on the honeycomb lattice, a ring polymer at the collapse point has the statistics of a percolation cluster boundary, i.e. a loop in the honeycomb lattice loop model at $n = 1, x = 1$ (see Sects. 1.2 and 1.5). We will refer to the associated RG fixed point as the Duplantier Saleur fixed point. The loop models of Chap. 2 with $n = 1$, and tricolour percolation [33–35], yield examples of smart walks in 3D. Such models serve simultaneously as models for polymers on the verge of collapse, for deterministic walks in a random environment (Lorentz lattice gases), and for line defects in disordered systems.

The key point, from the point of view of field theory and the classification of polymer models, is that models for the Θ point which map to smart walks have an enhanced global symmetry. This is simply because they can be formulated as loop ensembles at fugacity $n = 1$ rather than at fugacity $n' \to 0$. We have seen this above for the ISAT, and a similar phenomenon occurs for models without crossings.

In particular, the Duplantier Saleur Θ point, being described by the CP^{n-1} model with $n \to 1$, has a global $SU(n \to 1)$ symmetry. Modifications to the Boltzmann weight can break this symmetry in various ways, as we will discuss elsewhere [36]. Perturbations which allow crossings while *preserving* the smart walk property introduce a relevant symmetry-breaking operator that gives a crossover to the ISAT Θ point, with $SO(n \to 1)$ symmetry. However, preserving the smart walk property requires fine-tuning. More generic perturbations introduce an additional relevant perturbation to the CP^{n-1} Lagrangian, reducing the symmetry to $O(n')$,[6] and the RG flow (of the model at its collapse transition) no longer leads to the ISAT Θ point. Instead we would expect to see a new fixed point with exponents different from both the Duplantier Saleur model and the ISAT. (Of course, subtleties are always possible—e.g. the RG flow diagram could in principle involve two fixed points with the same exponents!) We would expect this more stable fixed point to correspond to the tricritical $O(n' \to 0)$ model; this is also consistent with a naive analysis of perturbations to the ISAT fixed point, as mentioned in Sect. 6.6.

6.8 Note on Three Dimensions

The approach to the ISAT described in this chapter generalises simply to the 3D ISAT model. For a certain choice of interactions, the model at its collapse point is related to a smart walk. A mapping similar to that described in Sect. 6.3 is possible, and shows that this collapse point is again an infinite order multicritical point, governed by the Goldstone phase of the $O(n \to 1)$ model. This explains the absence [25, 26] of the logarithmic corrections, for example to the radius of gyration, expected for a generic model (which would be described by the tricritical $O(n \to 0)$ model at its upper critical dimension [2]).

6.9 Conclusion

Above we gave a field theoretic treatment of the interacting self-avoiding trail that explains its phase structure and critical behaviour, and shows that the collapse transition in this model is an infinite order multicritical point, despite the fact that the mapping to the CPLC takes it to a point in the middle of a phase.

---

[6] Here we are referring to generic perturbations that introduce crossings. However, even if the perturbation does not introduce crossings, the counting of relevant operators suggests that the CP^0 fixed point will still generically be destabilised [36]. It is not clear how to reconcile this with the numerical results mentioned above [11, 12].

We also briefly touched on the classification of more general models for polymer collapse in two dimensions, and reached the surprising conclusion that critical exponents for a fully generic model (in which crossings are not forbidden and the Boltzmann weight is not fine-tuned to give a smart model) may be different from those of any of the well-known models for which exact results are available. However, various subtleties could arise, and it should be noted that the numerical simulations of Ref. [30], which considered a model that appears to be generic by our criteria, gave estimates for the thermal exponent ν in the range 0.55–0.57, roughly compatible with the Duplantier Saleur value $4/7 \simeq 0.571$. It is remarkable that there is still an open question here: further numerical work in this area would clearly be desirable.

There are several other outstanding questions relating to polymer collapse in 2D. One is to do with the exactly solved model without crossings mentioned briefly in Sect. 6.3, for which it is unclear how to reconcile numerical and analytical results [4, 27, 37, 38]. Another concerns a series of multicritical points found in certain supersymmetric theories [39, 40], with exponents that coincide with those in the approximate Flory argument for multicritical polymers [3]—it is not clear to what, physically, this series of multicritical points corresponds.

References

1. P.G. de Gennes, Phys. Lett. A **38**, 339 (1972)
2. P.G. de Gennes, J. de Physique Lettres **36**, L55 (1975)
3. J. Cardy, *Scaling and Renormalization in Statistical Physics* (Cambridge University Press, Cambridge, 1996)
4. H.W.J. Blöte, B. Nienhuis, J. Phys. A: Math. Gen. **22**, 1415 (1989)
5. B. Duplantier, H. Saleur, Phys. Rev. Lett. **59**, 539 (1987)
6. A.L. Owczarek, T. Prellberg, J. Stat. Phys. **79**, 951 (1995)
7. D.P. Foster, J. Phys. A: Math. Theor. **42**, 372002 (2009)
8. J. Doukas, A.L. Owczarek, T. Prellberg, Phys. Rev. E **82**, 031103 (2010)
9. W. Guo, H.J. Blöte, B. Nienhuis, Int. J. Mod. Phys C **10**, 301 (1999)
10. B. Duplantier, H. Saleur, Phys. Rev. Lett. **62**, 1368 (1989)
11. P. Grassberger, R. Hegger, J. Phys. I France **5**, 597 (1995)
12. S. Caracciolo, M. Gherardio, M. Papinutto, A. Pelissetto, J. Phys. A: Math. Theor. **44**, 115004 (2011)
13. T. Prellberg, A.L. Owczarek, J. Phys. A: Math. Gen. **27**, 1811 (1994)
14. A. Coniglio, N. Jan, I. Majid, H.E. Stanley, Phys. Rev. B **35**, 3617 (1987)
15. J.W. Lyklema, J. Phys. A. Math. Gen. **18**, L617 (1985)
16. A. Guha, H.A. Lim, Y. Shapir, J. Phys. A: Math. Gen. **21**, 1043 (1988)
17. H. Meirovitch, H.A. Lim, Phys. Rev. A **38**, 1670 (1988)
18. Y. Shapir, Y. Oono, J. Phys. A: Math. Gen. **17**, L39 (2984)
19. A. Bedini, A.L. Owczarek, T. Prellberg, Physica A **392**, 1602 (2013)
20. A. Bedini, A.L. Owczarek, T. Prellberg, Phys. Rev. E **87**, 012142 (2013)
21. H.W.J. Blöte, M.T. Batchelor, B. Nienhuis, Physica A **251**, 95 (1998)
22. J.L. Jacobsen, J. Kondev, J. Stat. Phys **96**, 21 (1999)
23. D.P. Foster, Phys. Rev. E. **84**, 032102 (2011)
24. D.P. Foster, J. Phys. A: Math. Theor. **43**, 335004 (2010)

25. T. Prellberg, A.L. Owczarek, Phys. Rev. E **51**, 2142 (1995)
26. A. Bedini, A.L. Owczarek, T. Prellberg, Phys. Rev. E **86**, 011123 (2012)
27. S.O. Warnaar, M.T. Batchelor, B. Nienhuis, J. Phys. A: Math. Gen. **25**, 3077 (1992)
28. R.M. Ziff, X.P. Kong, E.G.D. Cohen, Phys. Rev. A **44**, 2410 (1991)
29. J.L. Jacobsen, N. Read, H. Saleur, Phys. Rev. Lett. **90**, 090601 (2003)
30. A. Bedini, A.L. Owczarek, T. Prellberg, J. Phys. A. **46**, 085001 (2013)
31. A. Bedini, A.L. Owczarek, T. Prellberg, J. Phys. A: Math. Theor. **47**, 145002 (2014)
32. A.D. Sokal, Nucl. Phys. B (Proc. Suppl.) **47**, 172 (1996)
33. R.M. Bradley, P.N. Strenski, J.-M. Debierre, Phys. Rev. A **45**, 8513 (1992)
34. R.M. Bradley, J.-M. Debierre, P.N. Strenski, Phys. Rev. Lett **68**, 2332 (1992)
35. R.M. Bradley, J.-M. Debierre, P.N. Strenski, J. Phys. A **25**, L541 (1992)
36. A. Nahum, RG flows for Θ-point polymers, in preparation
37. D.P. Foster, C. Pinettes, J. Phys. A: Math. Theor. **45**, 505003 (2012)
38. A. Bedini, A.L. Owczarek, T. Prellberg, J. Phys. A: Math. Theor. **46**, 265003 (2013)
39. H. Saleur, Nucl. Phys. B **382**, 532 (1992)
40. J. Cardy, J. Phys. A: Math. Gen. **34**, L665 (2001)

Chapter 7
Outlook

We briefly recall some of the topics we have discussed, noting a few open questions.

The loop models of Chap. 2 proved a useful starting point for investigating general 3D loop ensembles, and also led us to open questions about critical points in CP^{n-1} or NCCP^{n-1} field theories and $SU(n)$–symmetric quantum magnets.

For the CP^{n-1} model, the RG account of Sect. 2.8.2 explains the apparent critical behaviour in the loop models with $n = 3$ and also in the bilayer quantum magnets of Ref. [1]. A question that is still open is whether this is truly a critical point or whether we just have a correlation length rendered 'exponentially' large by the close proximity of n_c. Another interesting outcome of the RG treatment is the possibility of multiple upper critical dimensions for the same field theory. This claim remains to be tested numerically.

A more pressing question concerns the peculiar behaviour of the 3D NCCP^{n-1} model close to $n = 2$. Early work on deconfined criticality [2] used the fact that the NCCP^{n-1} model has continuous transitions both at $n = 1$ and for large n (as we see from the duality with the XY model and from a saddle-point analysis respectively) to argue that the simplest hypothesis was a continuous transition at the intermediate value $n = 2$. But in view of the large corrections to scaling seen in simulations, we may question whether the $n = \infty$ critical point is analytically connected to that at $n = 1$, or whether something 'drastic' happens to the RG flows at a value of n near two. It may be possible to find an approach to this problem analogous to that for the compact case in Sect. 2.8.2. We have touched fleetingly on some issues that may be relevant to this puzzle in Chaps. 2 and 4; other useful pieces of information may come from the $2 + \epsilon$ expansion for the CP^{n-1} sigma model and the $4 - \epsilon$ expansion in the Abelian Higgs formulation. Conceivably, simulations at noninteger n slightly above and slightly below two could also be useful, for example in the deconfined model described in Sects. 3.3 and 3.4.

The extended phases in the 3D loop ensembles in Chaps. 2 and 4 also turned out to have some interesting features, despite their simplicity from the point of view of field theory. The basic feature—Brownian exponents—follows almost trivially from the replica-augmented field theory, while previously it had to be argued for in an

A. Nahum, *Critical Phenomena in Loop Models*, Springer Theses,
DOI 10.1007/978-3-319-06407-9_7

ad hoc manner. The replica trick may also be used to explain the Poisson Dirichlet distribution of loop lengths. This distribution answers the natural question of what the effect of the value of n is in the extended phase.

The fractal geometry of vortex lines may not be a topic of pressing practical importance,[1] but from a more abstract point of view the universality classes identified in Chap. 4 play an important role in the classification of critical disordered systems in three dimensions.

Turning to two dimensions, the critical points in Chap. 5 challenge us to find techniques to calculate their properties. For example, are there alternatives to the RP^{n-1} description? They also pose questions about what the space of critical points looks like for 2D loop models once we allow crossings.

As noted in the conclusion to Chap. 5, another interesting set of questions is to do with the relationship (or perhaps only analogy) with localisation. In fact, the relationship between loops and localisation is an interesting one more generally. In 1985 Anderson asked about the quantum Hall transition: 'Are we sure that this is not a case where, unlike the standard one [...], the crossover is from localization to percolation and not vice versa?' [4]. We now know, of course, that the RG flow is from percolation to the distinct universality class of the quantum Hall transition. However, an understanding of this as an RG flow in an effective field theory is missing, and this is a fundamental gap in our understanding of disordered systems. We have made initial progress in this direction and hope to report on it soon.

A more abstruse question about the 2D RP^{n-1} models and their supersymmetric extensions is whether they have natural geometrical descriptions that are 'dual' (in a loose sense) to those discussed in Chap. 5. These would be descriptions in which the geometric objects are built out of worldlines of the gauge-invariant field Q rather than worldlines of $\mathbf{S}$ (cf. the discussion in Appendix. A). The case $n = 2$ is a simple one: both worldlines of $\mathbf{S}$ and worldlines of Q form ensembles of loops with fugacity two (this is analogous to the discussion in Sect. 2.12 about the CP^1 model). It would be interesting to know whether there is any simple geometrical relationship between these two ensembles, analogous to the SLE duality mentioned in Sect. 1.4.3.

For $n = 2$, the critical behaviour is more conventional than for $n < 2$, and is governed by the XY model, at least for properties which do not require the replica trick. However, there are some unusual features. In particular, it should be possible to see a continuously varying fractal dimension in the loop models with crossings at $n = 2$, as a result of the critical phase of the XY model (which is what replaces the Goldstone phase when $n = 2$). By considering the 2-leg operator at $n = 2$, the fractal dimension is

---

[1] We can of course imagine toy models in which vortex geometry determines physical observables. For example, consider a superconductor with topologically protected low energy states in the cores of vortices of the gap function [3]. In an appropriate semiclassical limit, taking the gap function as a quenched random field with bias, we can obtain a localisation transition that by construction is in the oriented vortex universality class of Chap. 4. This is not intended as a realistic model, but amusingly it captures some of the *exact* critical exponents for localisation in class C superconductors as a result of the mapping from localisation in class C to completely packed loops.

$$d_f = 2 - \frac{1}{\pi K} \tag{7.1}$$

in the normalisation of Chap. 5. The minimal fractal dimension is determined by the critical value of K at which the Kosterlitz Thouless transition occurs. However the KT transition can occur by a proliferation of either single or double vortices in Q (recall that at $n = 2$ the vortices have at $\mathbb{Z}$-valued charge, unlike for $n < 2$). The former are forbidden on the line $q = 1/2$, leading to a larger critical stiffness: this gives $d_f \geq 3/2$. This bound is the fractal dimension in the critical loop model without crossings. For $q \neq 1/2$, we have instead[2] $d_f \geq 15/8$.

Chapter 6 addressed some longstanding questions about Θ point polymers. We will not repeat the summary given a few pages ago, but we note that plenty of concrete questions for the ISAT could be tackled using the field theory of Chap. 6.

The idea of a 'smart walk' discussed in Chap. 6 generalises to 'smart surfaces' [5]. A critical percolation cluster boundary in 3D provides a 'smart' model for a self-avoiding surface, of fluctuating topology, with certain attractive interactions. One may write replica-like lattice field theories (see e.g. [6]) for a self avoiding surface—these include lattice gauge fields, but the replica-like symmetry is simply S_q (permutational) with a $q \to 0$ limit to extract a single surface. On the other hand the 'smart surface' is associated with the $q \to 1$ limit familiar from percolation. This is in analogy to the symmetry enhancement we saw for polymers. Perhaps we can determine how the smart surface fits into a more general phase diagram for a random surface (how many parameters need to be tuned to reach it) by considering symmetry-breaking perturbations of the $(q \to 1)$-state Potts model.

For the future, generalisations of the quantum loop models mentioned in Sect. 1.6 are an intriguing direction, promising further gapped and gapless states with subtle connections to random geometry.

References

1. R.K. Kaul, Phys. Rev. B **85**, 180411 (2012)
2. T. Senthil, L. Balents, S. Sachdev, A. Vishwanath, M.P.A. Fisher, Phys. Rev. B **70**, 144407 (2004)
3. J.C.Y. Teo, C.L. Kane, Phys. Rev. B **82**, 115120 (2010)
4. P.W. Anderson, in *Localization, Interaction and Transport Phenomena*, ed. by B. Kramer, Y. Bruynserade (Springer, Berlin, 1985)
5. R.M. Bradley, P.N. Strenski, J.-M. Debierre, Phys. Rev. B **44**, 76 (1991)
6. P.D. Gujrati, Phys. Rev. B **39**, 2494 (1989)

[2] As $n \to 2$, the fractal dimension at the RP^{n-1} critical point tends to this value (15/8).

Appendix A
Potts Domain Walls and CP^{n-1}

At the end of Chap. 2, we speculated about geometrical ensembles made up of worldlines of the field Q, rather than worldlines of $\mathbf{z}$. Consider the two-dimensional case, when the CP^{n-1} model can be made critical by the addition of a θ term with $\theta = \pi$:

$$\mathcal{L} = \frac{K}{4} \operatorname{tr} (\nabla Q)^2 + \frac{\theta}{2\pi} \epsilon_{\mu\nu} \operatorname{tr} Q \nabla_\mu Q \nabla_\nu Q. \tag{A.1}$$

The loop model on the L lattice (pictured in Fig. 2.1) provides a lattice regularisation of the ensemble of $\mathbf{z}$ worldlines in this field theory. On the other hand, we know that the partition function Z of this loop model can be re-expressed as the partition function of the n^2-state Potts model on a dual lattice, using the Fortuin-Kasteleyn representation of the latter. In turn the Potts partition function can be expressed as a sum over domain wall configurations. This yields a second graphical expansion for Z in terms of branching nets; these are of course worldlines of Q. (See Refs. [1, 2] for more sophisticated analyses of the relation between these two kinds of geometrical objects and the Temperley-Lieb and Murakami-Birman-Wenzl algebras.)

We show here that there is a simple lattice regularisation of the field theory (A.1) whose strong coupling expansion directly generates Potts domain walls. There are outstanding questions about the random geometry of these objects [3–6]. Natural geometrical correlation functions have recently been defined and, on the basis of numerical work, conjectures have been made for their exact scaling dimensions [3, 4]. It is not known how to relate these conjectures to field theory. At present it is not clear whether the following construction can shed light on these issues—i.e. whether these correlation functions can be expressed using the lattice model below and an appropriate replica limit—but this is a motivation for the construction. It also yields what is perhaps the most direct way of seeing the relation between the Potts model and the CP^{n-1} model.

We situate our CP^{n-1} spins Q_l on the *links* l of the honeycomb lattice. Microscopically, they are defined as usual by $Q_l = \mathbf{z}_l \mathbf{z}_l^\dagger - 1$, with $|\mathbf{z}|^2 = n$. Let r denote a site of the honeycomb lattice, and in the following, let A, B, C denote the links surrounding r, labelled in anticlockwise order. Motivated by the form and the symmetries of

A. Nahum, *Critical Phenomena in Loop Models*, Springer Theses,
DOI 10.1007/978-3-319-06407-9

Fig. A.1 Possibilities for graphical expansion of (A.3) at a node

(A.1), we define the lattice partition function

$$Z = \mathrm{Tr} \prod_r \Big(1 + x \,\mathrm{tr}\,(Q_A Q_B + Q_B Q_C + Q_C Q_A) + y \,\mathrm{tr}\, Q_A Q_B Q_C \Big). \quad (A.2)$$

We will describe the case $n = 2$, and then briefly summarise the extension to $n \neq 2$ which is slightly more involved. For $n = 2$, write $Q = \sigma^i S^i$, with $S^2 = 1$:

$$Z = \mathrm{Tr} \prod_r \Big(1 + 2x\,(S_A.S_B + S_B.S_C + S_C.S_A) + 2iy\epsilon^{abc} S_A^a S_B^b S_C^c \Big). \quad (A.3)$$

This partition function has a simple graphical expansion. At each node we have a choice of five terms, leading to the graphical possibilities in Fig. A.1. We obtain a sum over configurations of nets, in which each line carries a flavour index $a = 1, 2, 3$ (the fact that flavour indices are conserved along the lines is of course a result of the single-link integral being proportional to a delta function, $\mathrm{Tr}\, S_l^a S_l^b = \delta^{ab}/3$). At a branching node (the rightmost possibility in Fig. A.1) the flavours of the three lines must be distinct because of the epsilon tensor in (A.3).

The weight of a given configuration depends on the total length of the lines and on the number of branching nodes. Additionally, each branching node contributes a complex weight which is i or $-i$ depending on whether the three flavours appear in anticlockwise or clockwise order around the node. However, a simple argument shows that these factors of i and $-i$ cancel. (It is enough to show that, for each loop made up of curves of alternating flavour 1 and 2, the number of clockwise nodes minus the number of anticlockwise nodes is a multiple of four.)

Finally, we map the net configurations to four-state Potts configurations. The Potts spins σ_F take the values **1**, **2**, **3**, **4**, and live on the *faces* F of the honeycomb lattice. Having fixed the Potts spin on one face, the others are fixed by a rule for how the colours are permuted when we cross a domain wall with flavour a:

$$a = 1: \;\; (\mathbf{1},\mathbf{2})(\mathbf{3},\mathbf{4}) \qquad a = 2: \;\; (\mathbf{1},\mathbf{4})(\mathbf{2},\mathbf{3}) \qquad a = 3: \;\; (\mathbf{1},\mathbf{3})(\mathbf{2},\mathbf{4}). \quad (A.4)$$

These rules are consistent, since these permutations are self-inverse (ensuring that we get back to the same colour if we cross a domain wall twice) and multiply to the identity (ensuring we get back to the same colour on encircling a branch point).

Altogether this correspondence yields

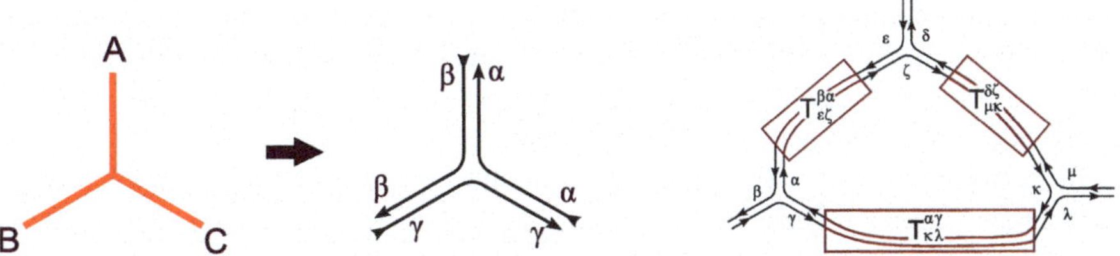

Fig. A.2 *Left* the term tr $Q_A Q_B Q_C = Q_A^{\alpha\beta} Q_B^{\beta\gamma} Q_C^{\gamma\alpha}$ means that a branching node emits *doubled lines* with indices as shown. *Right* a section of a diagram, after multiplying together the Ts on the links

$$Z \propto \sum_{\substack{\text{4-state}\\ \text{Potts configs}}} \left(\frac{2x}{3}\right)^{\text{length}} \left(\frac{y}{\sqrt{2}x^{3/2}}\right)^{\text{no. vortices}}. \tag{A.5}$$

Here 'length' is the total number of links covered by Potts domain walls, and a 'vortex' is a node where three distinct Potts colours meet. If we set $y = \sqrt{2}x^{3/2}$ and $2x/3 = e^{-J}$, we have the partition function for the 4-state Potts model on the triangular lattice with nearest-neighbour interactions:

$$Z \propto \sum_{\{\sigma\}} \exp\left(J \sum_{\langle FF'\rangle} \delta_{\sigma_F,\sigma_{F'}}\right). \tag{A.6}$$

We expect that when the parameters are such that the Potts model is critical, the continuum limit of the lattice field theory (A.3) is the $O(3)$ model (or CP^1 model), Eq. A.1, at $\theta = \pi$.

Now we summarise the extension to $n \neq 2$. We make a graphical expansion of (A.2) analogous to that above, which generates the same collection of nets. However, we no longer label the lines by a single conserved colour index. Each branching node emits labelled doubled lines, as shown in Fig. A.2, corresponding to the two indices on Q. These indices are not simply conserved along the lines, since the single-link integral is not just a product of delta functions:

$$\text{Tr } Q^{\beta\alpha} Q^{\alpha'\beta'} = \frac{n}{n+1} T^{\alpha\beta}_{\alpha'\beta'}, \quad T^{\alpha\beta}_{\alpha'\beta'} = \left(\delta_{\alpha\alpha'}\delta_{\beta\beta'} - \frac{1}{n}\delta_{\alpha\beta}\delta_{\alpha'\beta'}\right). \tag{A.7}$$

T should be regarded as a matrix, with rows and columns labelled by the upper and lower indices respectively. In the graphical expansion, each line carries a matrix product of Ts, one for each link. Since $T^2 = T$, this product simplifies, leaving one T for each line section (see Fig. A.2).

Finally, we must sum over the indices. This sum assigns a topology-dependent weight to each term in the graphical expansion. This weight turns out to correspond precisely to the number of configurations in the n^2-state Potts model that are compatible with a given domain wall configuration. This number is also known as the chromatic polynomial $P_G(n^2)$ of the graph G, which is defined so that its vertices are in correspondence with the Potts domains and its edges traverse the domain walls. Given n^2 colours, $P_G(n^2)$ is the number of ways of colouring the vertices of G so that no adjacent vertices have the same colour.

To perform the sum, we express the second equation in (A.7) graphically as $= \;-\frac{1}{n}$. Each domain wall configuration then reduces to a sum of diagrams made up of (single-line) loops; each such diagram is then weighted by a factor of n from the index sum on each loop. Fortunately, this sum of diagrams is the same sum that appears when we use a simple lemma—the deletion-contraction lemma—to calculate the chromatic polynomial $P_G(n^2)$. By considering the Euler character of the domain wall configuration, we may check that the factors of n work out correctly to give $P_G(n^2)$, up to a term that enters the 'vortex' fugacity. Altogether,

$$Z \propto \sum_{\substack{n^2\text{-state} \\ \text{Potts configs}}} \left(\frac{xn}{n+1}\right)^{\text{length}} \left(\frac{y}{x^{3/2}n^{1/2}}\right)^{\text{no. vortices}} . \tag{A.8}$$

For the choice $y = x^{3/2}n^{1/2}$, $xn/(n+1) = e^{-J}$, we again obtain the nearest-neighbour model (A.6), this time with n^2 states.

(As an aside, note that for sufficiently large y the transition is first order, and by varying both x and y we can access the tricritical Potts model. In the CP^{n-1} description (with $\theta = \pi$), the critical Potts model is the fixed point at large K which is stable in the K direction, and we expect that the tricritical Potts model corresponds to the fixed point at smaller K which is unstable in the K direction. Recalling the correspondence between CP^{n-1} and non-crossing loops, this is compatible with the known fact that while the critical Potts model corresponds to the dense phase, the tricritical Potts model corresponds to the dilute phase.)

References

1. P. Fendley and N. Read, J. Phys. A: Math. Gen. **35**, 10675 (2002).
2. P. Fendley and E. Fradkin, Phys. Rev. B **72**, 024412 (2005).
3. J. Dubail, J. L. Jacobsen, and H. Saleur, J. Phys. A: Math. Theor. **43**, 482002 (2010).
4. J. Dubail, J. L. Jacobsen, and H. Saleur, J. Stat. Mech. P12026 (2010).
5. A. Gamsa and J. Cardy, J. Stat. Mech. P08020 (2007).
6. J. Cardy, Nucl. Phys. B **565**, 506 (2000).

Appendix B
Phases for Hedgehogs & Vortices

B.1 Phases for Hedgehogs on the 3D L Lattice

Regarded as a lattice magnet for classical CP^{n-1} spins, the partition function (3.5) has the peculiar feature that although it is both local and gauge invariant, it is not simply expressed in terms of Q. The consequence of this, as noted in Sect. 3.3, is that the sigma model action arising from a derivative expansion may need to be supplemented by purely imaginary terms from nodes at which the phase of $\mathbf{z}$ changes abruptly. In the presence of hedgehogs such nodes are inevitable, even far from the hedgehog core. As a result the hedgehog fugacity acquires a spatially varying phase.

For simplicity we take the hedgehogs to be centred on a cube of C_1 or C_2—see Fig. 2.4, right. These locations form a bcc lattice. We take the origin at a bcc site, say the centre of the cube in Fig. 2.4, and the coordinate axes parallel to the links.

Focussing on the case $n = 2$ (the generalisation to other n is immediate), let us first consider the representative configuration in which the hedgehog is centred at the origin and the $O(3)$ spin S points in the radial direction. In polar coordinates, $0 \leq \theta \leq \pi$, $0 \leq \phi < 2\pi$, this is

$$S = (\sin\theta\cos\phi, \sin\theta\sin\phi, \cos\theta), \tag{B.1}$$

where the coordinates of a link are those of its midpoint. To write this in terms of $\mathbf{z}$ we must pick a gauge. It is convenient to choose one in which the Boltzmann weight

$$e^{-S_{\text{node}}} = \frac{1}{2}\left[(\mathbf{z}_o^\dagger \mathbf{z}_i)(\mathbf{z}_{o'}^\dagger \mathbf{z}_{i'}) + (\mathbf{z}_o^\dagger \mathbf{z}_{i'})(\mathbf{z}_{o'}^\dagger \mathbf{z}_i)\right] \tag{B.2}$$

is approximately one for as many nodes as possible. For the links with positive and negative z coordinate ($\theta < \pi/2$ and $\theta > \pi/2$) we take, respectively,

$$\mathbf{z} = (\cos\theta/2, e^{i\phi}\sin\theta/2) \qquad \text{and} \qquad \mathbf{z} = (e^{-i\phi}\cos\theta/2, \sin\theta/2). \tag{B.3}$$

A. Nahum, *Critical Phenomena in Loop Models*, Springer Theses,
DOI 10.1007/978-3-319-06407-9

We see from Fig. 2.4 that there are also links in the equatorial plane. For these we take

$$\mathbf{z} = \frac{1}{\sqrt{2}}(e^{-i\phi/2}, e^{i\phi/2}). \tag{B.4}$$

Now consider $e^{-S_{\text{node}}}$. In fact it will suffice to consider only nodes far from the core, and to treat $\mathbf{z}$ as constant except for the discontinuities in our gauge choice: other contributions to the action are either included in the spatially *in*dependent amplitude of the hedgehog fugacity, or are captured by the derivative expansion.

With the above gauge choice, the nodes at which $\mathbf{z}$ varies abruptly all lie in the equatorial plane, and have two of their links within this plane, one above it, and one below. Fig. 2.4 shows four such nodes (all yellow). From Eqs. B.2—B.4 we find that most of these nodes have $e^{-S_{\text{node}}} \simeq 1$, despite the variation in the phase of $\mathbf{z}$. However there is a string of nodes along the positive x axis ($\phi = 0$, $\theta = \pi/2$) that contribute minus signs to the Boltzmann weight ($e^{-S_{\text{node}}} \simeq -1$). If we translate the core of the hedgehog by the vector (2, 0, 0), while keeping the configuration far from the core fixed, we change number of nodes on this string by one. Therefore this translation changes the sign of the Boltzmann weight.

Let the phase term in the hedgehog fugacity be denoted $e^{i\eta(r)}$, where the spatial vector r lies on a bcc site. (An 'antihedgehog' of negative topological charge has phase factor $e^{-i\eta(r)}$, as we see from Eq. B.2 and the fact that complex conjugating $\mathbf{z}$ exchanges hedgehogs and antihedgehogs.) The phase $\eta(r)$ is defined only up to a constant—for example in a closed system there are equal numbers of hedgehogs and antihedgehogs, so the constant part of η drops out.

It may be seen from Fig. 2.4 that the translational symmetry between bcc sites is not spoiled by the link orientations. Using this, we may argue that $\eta(r)$ is of the form $\eta(r) = k.r$ for some momentum k. By the above calculation, $e^{i\eta(r)} = -e^{i\eta(r+2\hat{x})}$, where $\hat{x} = (1, 0, 0)$. By symmetry, we have similar results for translations in the y and z directions. This is enough to fix k up to a sign:

$$k = \pm\frac{\pi}{2}(1, 1, 1). \tag{B.5}$$

One of these signs applies to the hedgehog and one to the anti-hedgehog. We have not fixed which is which, but it does not matter.

With this k, the hedgehog fugacity takes four distinct values on the four sublattices of the bcc lattice, proportional to ± 1 and $\pm i$. This is the result quoted in Chap. 3.

B.2 Phases for $\mathbb{Z}_2$ Vortices in 2D

A similar but even simpler calculation applies to the completely-packed loops with crossings on the square lattice (CPLC) considered in Chap. 5. As described there, the spin is a point on RP^{n-1}, parameterised by a vector $\mathbf{S}$ with $\mathbf{S} \sim -\mathbf{S}$. An example of a $\mathbb{Z}_2$ vortex, in polar coordinates, is $\mathbf{S} = (\cos\theta/2, \sin\theta/2)$. We take the origin at the centre of a plaquette. In this gauge, the links at $y = 0, x > 0$ have $\mathbf{S} = (1, 0)$, while far from the origin the links just below the positive x axis have $\mathbf{S} \simeq -(1, 0)$. The lattice Boltzmann weight (5.22) shows that all the nodes at $y = -1/2$, $x > 0$ contribute minus signs, in analogy to the string of negative nodes in the 3D example above. Similar reasoning shows that the vortex fugacity has opposite signs for plaquettes of each sublattice. Therefore the coarse-grained vortex fugacity vanishes at $q = 1/2$, regardless of p, since symmetry ensures the amplitude of the microscopic fugacity is the same on each sublattice. When $q \neq 1/2$, this symmetry is broken, spoiling the cancellation, and the coarse-grained fugacity is proportional to $(q - 1/2)$ as we found using the reduction from the CP^{n-1} model.

The manufacturer's authorised representative in the EU is Springer Nature Customer Service Centre GmbH, Europaplatz 3, 69115 Heidelberg, Germany. If you have any concerns regarding our products, please contact ProductSafety@springernature.com

Printed and bound by CPI Group (UK) Ltd, Croydon, CR0 4YY
15/07/2026
02167645-0010